Lecture Notes in Mathematics

Edited by J.-M. Morel, F. Takens and B. Teissier

Editorial Policy
for the publication of monographs

1. Lecture Notes aim to report new developments in all areas of mathematics – quickly, informally and at a high level. Monograph manuscripts should be reasonably self-contained and rounded off. Thus they may, and often will, present not only results of the author but also related work by other people. They may be based on specialized lecture courses. Furthermore, the manuscripts should provide sufficient motivation, examples and applications. This clearly distinguishes Lecture Notes from journal articles or technical reports which normally are very concise. Articles intended for a journal but too long to be accepted by most journals, usually do not have this "lecture notes" character. For similar reasons it is unusual for doctoral theses to be accepted for the Lecture Notes series.

2. Manuscripts should be submitted (preferably in duplicate) either to one of the series editors or to Springer-Verlag, Heidelberg. In general, manuscripts will be sent out to 2 external referees for evaluation. If a decision cannot yet be reached on the basis of the first 2 reports, further referees may be contacted: the author will be informed of this. A final decision to publish can be made only on the basis of the complete manuscript, however a refereeing process leading to a preliminary decision can be based on a pre-final or incomplete manuscript. The strict minimum amount of material that will be considered should include a detailed outline describing the planned contents of each chapter, a bibliography and several sample chapters.
Authors should be aware that incomplete or insufficiently close to final manuscripts almost always result in longer refereeing times and nevertheless unclear referees' recommendations, making further refereeing of a final draft necessary.
Authors should also be aware that parallel submission of their manuscript to another publisher while under consideration for LNM will in general lead to immediate rejection.

3. Manuscripts should in general be submitted in English.
Final manuscripts should contain at least 100 pages of mathematical text and should include
- a table of contents;
- an informative introduction, with adequate motivation and perhaps some historical remarks: it should be accessible to a reader not intimately familiar with the topic treated;
- a subject index: as a rule this is genuinely helpful for the reader.

Continued on inside back-cover

Lecture Notes in Mathematics 1759

Springer
Berlin
Heidelberg
New York
Barcelona
Hong Kong
London
Milan
Paris
Singapore
Tokyo

Yukitaka Abe Klaus Kopfermann

Toroidal Groups

Line Bundles, Cohomology and Quasi-Abelian Varieties

Springer

Authors

Yukitaka Abe
Department of Mathematics
Faculty of Science
Toyama University
Gofuku 3190
Toyama 930-8555, Japan

E-mail: abe@sci.toyama-u.ac.jp

Klaus Kopfermann
Institut für Mathematik
Universität Hannover
Welfengarten 1
30167 Hannover, Germany

E-mail: kostock@inet.co.th

Cataloging-in-Publication Data applied for

Die Deutsche Bibliothek - CIP-Einheitsaufnahme

Abe, Yukitaka:
Toroidal groups : line bundles, cohomology and quasi-Abelian varieties / Yukitaka Abe ; Klaus Kopfermann. - Berlin ; Heidelberg ; New York ; Barcelona ; Hong Kong ; London ; Milan ; Paris ; Singapore ; Tokyo : Springer, 2001
(Lecture notes in mathematics ; 1759)
ISBN 3-540-41989-6

Mathematics Subject Classification (2000): 22E10, 22E99

ISSN 0075-8434
ISBN 3-540-41989-6 Springer-Verlag Berlin Heidelberg New York

Springer-Verlag Berlin Heidelberg New York
a member of BertelsmannSpringer Science+Business Media GmbH

http://www.springer.de

Printed in Germany

Typesetting: Camera-ready TEX output by the authors
SPIN: 10836500 41/3142-543210 - Printed on acid-free paper

Preface

Toroidal groups are the missing links between torus groups and any complex Lie groups. Many phenomena of complex Lie groups such as pseudoconvexity and the structure of cohomology groups can be understood only through the concept of toroidal groups. The different behavior of the cohomology groups of complex Lie groups can be characterized by the properties of their toroidal groups appearing in their centers.

Toroidal groups have not been treated systematically in a book. So the oldest living mathematician who worked in this field and the youngest working in it decided to give a comprehensive survey about the main results concerning these groups and to discuss open problems.

Toroidal groups are the non–compact generalization of the torus groups. As complex manifolds they are convex in the sense of Andreotti and Grauert. As complex Lie groups some of them have a similar behavior to complex tori, others are different with for example non–Hausdorff cohomology groups, whence completely new methods must be used.

The aim of these lecture notes is to describe the fundamental properties of toroidal groups. As a result of the meromorphic reduction theorem the quasi–Abelian varieties are of special interest. Their basic description ends in the third chapter with the Main Theorem.

This theory - in honour of SOPHUS LIE - was introduced to a wide public at the Conference "100 Years after Sophus Lie" in Leipzig, on July 8-9, 1999.

The first-named author wishes to thank the ALEXANDER VON HUMBOLDT FOUNDATION for partial support.

Hannover and Toyama, December 1998

Yukitaka Abe and Klaus Kopfermann

Symbols

$\lvert z \rvert$, $\lVert z \rVert$	euclidian norm, maximum norm	
$\lvert\sigma\rvert$	$:= \sum_j \lvert\sigma_j\rvert$ for $\sigma \in \mathbf{Z}^n$	
$\lvert J\rvert$	:= p for the multiindex $J = (j_1, \cdots, j_p)$	
$\langle z, w\rangle$	$:= \sum_j z_j w_j$, the skalar product of $z, w \in \mathbf{C}^n$	
$\mathbf{e}(z)$	$:= (\exp(2\pi i z_1), \cdots, \exp(2\pi i z_n))$ for $z \in \mathbf{C}^n$	
$\mathrm{diag}(d_1, \cdots, d_n)$	diagonal matrix	
Re, Im	real-, imaginary part of a complex number	
$\mathbf{P}_n$	n–dimensional complex projective space	
$\mathcal{H}(X)$, $\mathcal{M}(X)$	holomorphic, meromorphic functions on X	
ker, im	kernel, image of a map	
Ker$H$	kernel of a bilinear or Hermitian form H	69
Λ, Γ	lattices = discrete subgroups of a **C**–linear space	4
$\mathbf{R}_\Lambda, \mathbf{C}_\Lambda$	the **R**-span, **C**-span of Λ	4
MC_Λ	the maximal **C**–linear subspace of $\mathbf{R}_\Lambda$	4
$\mathbf{R}_\Lambda/\Lambda$	the maximal real subtorus	4
$\mathrm{MC}_\Lambda/(\mathrm{MC}_\Lambda \cap \Lambda)$	the maximal complex subgroup of $\mathbf{R}_\Lambda/\Lambda$	4
(I)	Irrationality conditions for toroidal groups	6,10,11
(TS), (TT)	conditions for theta groups	44
(PRC) -H) -S)	period relations	61,62
α_λ, β_λ	automorphic factors	25,29,57
ϑ_λ, ι_λ	theta factor, wild factor	37,60
$\mathcal{A}_\alpha$	automorphic forms belonging to α	27
$c_1(L)$	Chern class of the line bundle L	26
Pic(X)	Picard group of all line bundles of X	26
$\mathrm{Pic}_0(X)$	topological trivial line bundles of X	40
NS(X)	Néron-Severi group of X	43
Div(X)	group of all divisors on X	
$\mathcal{O}, \Omega^m$	sheaf of germs of hol. functions, m–forms	26,47
$\mathcal{E}, \mathcal{E}^r, \mathcal{E}^{m,p}$	sheaf of germs of $\mathcal{C}^\infty$–funct., r–forms, (m,p)–forms	47
$\mathcal{F}$	sheaf of germs of $\mathcal{C}^\infty$ functions, hol. along the fibre	49
$\mathcal{F}^{m,p}$	sheaf of germs of $\mathcal{C}^\infty$ (m,p)–forms with coeff. in $\mathcal{F}$	49
$Av(\theta)$	average of a differential form	75
T_K	tangent bundle of K	74
$H_1 \simeq_\Lambda H_2$	Λ–equivalent Hermitian forms	84
H_Λ, A_Λ	restr. of H to $\mathrm{MC}_\Lambda \times \mathrm{MC}_\Lambda$ and $A := \mathrm{Im}H$ to $\mathbf{R}_\Lambda \times \mathbf{R}_\Lambda$	88
(C0)~(C4)	conditions on Riemann forms	93-100
$\mathcal{H}_n$	Siegel upper half space	113

Contents

Introduction

P. COUSIN studied in 1910 meromorphic functions of two variables with triple periods and discovered a group which does not contain **C** or $\mathbf{C}^*$ as direct summand. Such a group is a toroidal group.

Toroidal groups X have maximal closed Stein subgroups $N \simeq C^{*m}$, such that the quotient X/N is a complex torus and X a C^{*m}–fibre bundle over X/N. The fact that such a bundle can be described by an exponential systems of an automorphic factor led to many results.

K. KOPFERMANN studied in 1964 systematically toroidal groups defined by an irrationality condition of a lattice Λ. He proved the characteristic and the Hermitian decomposition of theta factors for further results and gave an example of a non–compact toroidal group without meromorphic functions.

A. MORIMOTO proved in 1965 that toroidal groups appear as Steinizers of connected complex Lie groups and proved the holomorphic reduction theorem for complex Lie groups.

F. GHERARDELLI and A. ANDREOTTI defined in 1971–73 quasi–Abelian varieties and proved some of their properties.

H. KAZAMA proved in 1973 that toroidal groups are q–complete and therefore pseudoconvex.

A. HEFEZ proved in 1979 that the meromorphic function fields on non–compact quasi–Abelian varieties have an infinite transcendence degree over **C**. Also POTHERING obtained this theorem in 1977.

CHR. VOGT continued the systematic work on toroidal groups in 1982 by studying line bundles on toroidal groups. He proved that every holomorphic line bundle over a toroidal group X is defined by a theta factor, if the cohomology group $H^1(X, \mathcal{O})$ has finite dimension. He found in 1983 general results about cohomology groups of toroidal theta groups.

H. KAZAMA analysed in 1984 the cohomology groups of all toroidal groups. He found final results about the Dolbeault cohomology of all complex Lie groups by studying their maximal toroidal subgroups. Together with T. UMENO he created a cohomology theory of holomorphic fibre spaces.

Y. ABE systematically characterized in 1987-89 toroidal groups with positive line bundles, recovered earlier partial results as special cases in his Main Theorem about these quasi–Abelian varieties and proved that these varieties are analogous to the Abelian varieties. He also proved the meromorphic reduction theorem for toroidal groups.

F. CAPOCASA and F. CATANESE added in 1991 the classical aspect of the existence of a non–degenerate meromorphic function.

To prove the *existence of automorphic forms* belonging to a given automorphic factor or equivalently of non–trivial *sections in line bundles* defined by the same automorphic factor is the fundamental step to the construction of meromorphic functions and embeddings into complex projective spaces.

P. COUSIN proved in 1910 that a topologically trivial line bundle[1] on a toroidal group of rank n+1 has non–trivial sections, if and only if it is analytical trivial.
Y. ABE proved in 1995 that the assumption can be generalized to line bundles with Hermitian forms vanishing on the maximal complex linear subspace. There exists a topologically non–trivial theta bundle which satisfies this condition.
Y. ABE, later F. CAPOCASA and F. CATANESE and recently S. TAKAYAMA continued to study meromorphic functions on quasi–Abelian varieties and mappings to complex projective spaces by studying sections in line bundles.
More recently, Y. ABE is trying explicitly to construct automorphic forms in the general case. He and also F. CATANESE are interested in an extension of MUMFORD's index theorem.

[1] In present day language

1. The Concept of Toroidal Groups

The general concept of toroidal groups was introduced in 1964 by KOPFERMANN *by the irrationality condition.* MORIMOTO *considered in 1965 complex Lie groups which lack non–constant holomorphic functions and contributed basic properties of them.* KAZAMA *continued the work with pseudoconvexity and cohomology groups.*

1.1 Irrationality and toroidal coordinates

The fundamental tool are toroidal coordinates which allow to select toroidal groups out of others by irrationality conditions. Toroidal groups can be represented as principal fibre bundles over a complex torus group with a Stein fibre isomorphic to a $\mathbf{C}^{*m}$.

Toroidal groups

The concept of complex torus groups leads to

1.1.1 Definition

A **toroidal group** is an Abelian complex Lie group on which every holomorphic function is constant.

Toroidal groups have several names in literature such as (H,C)-*groups*, that means all holomorphic functions are constant, or *quasi-torus*[1]. Sometimes a quasi-torus is any connected Abelian Lie group.

A consequence of a theorem of MORIMOTO is that every complex Lie group on which all holomorphic functions are constant is Abelian (Remark 1.2.3 on p 18).

$\mathbf{R}^n$ is the unique connected and Abelian real Lie group of dimension n which is simply connected and the $\mathbf{C}^n$ the unique connected and Abelian complex Lie group of complex dimension n which is simply connected.

[1] Also called Cousin quasi-torus, because COUSIN had an example of such a group (see p 1)

1.1.2 Proposition
Every connected Abelian complex Lie group is isomorphic as complex Lie group to $\mathbf{C}^n/\Lambda$ where Λ is a discrete subgroup of $\mathbf{C}^n$.

Proof
If such a Lie group X has the complex dimension n, then $\mathbf{C}^n$ is its universal covering group with projection $\pi : \mathbf{C}^n \to X$ which is a complex homomorphism. Therefore $\Lambda := \ker \pi \simeq \pi_1(X)$ is a discrete subgroup of $\mathbf{C}^n$ such that $X \simeq \mathbf{C}^n/\Lambda$.
Q.E.D.

A **lattice** $\Lambda \subset \mathbf{R}^m$ is a discrete subgroup of $\mathbf{R}^m$. A lattice $\Lambda \subset \mathbf{C}^n$ **represents** the Abelian complex Lie group $X := \mathbf{C}^n/\Lambda$. A **basis** of a lattice $\Lambda \subset \mathbf{C}^n$ is an ordered set or matrix $P = (\lambda_1, \cdots, \lambda_r)$ of $\mathbf{R}$–independent $\mathbf{Z}$-generators of Λ.

For a lattice $\Lambda \subset \mathbf{C}^n$ with basis $(\lambda_1, \cdots, \lambda_r)$ let

$$\mathbf{R}_\Lambda := \{x_1\lambda_1 + \cdots + x_r\lambda_r : \; x_1 \in \mathbf{R}, \cdots, x_r \in \mathbf{R}\}$$

be the $\mathbf{R}$–span of Λ.

The complex rank of a basis P is said to be the complex rank of a lattice Λ. If the complex rank of a lattice $\Lambda \subset \mathbf{C}^n$ is $m < n$, then after a linear change of the coordinates we can assume that the $\mathbf{C}$–span $\mathbf{C}_\Lambda := \mathbf{R}_\Lambda + i\mathbf{R}_\Lambda$ of Λ is the subspace of the first m coordinates. If the complex and real rank of $\Lambda \subset \mathbf{C}^n$ are n, then after a linear change of the coordinates we can assume that $e_1, \cdots, e_n$ are $\mathbf{Z}$–generators of Λ so that $\mathbf{C}^n/\Lambda = \mathbf{C}^n/\mathbf{Z}^n \simeq (\mathbf{C}/\mathbf{Z})^n \simeq \mathbf{C}^{*n}$ by **exponential map**

$$\mathbf{e}(z) := (\exp(2\pi i z_1), \cdots, \exp(2\pi i z_n)) \quad (z \in \mathbf{C}^n),$$

where $\mathbf{C}^*$ is the multiplicative group of the complex numbers.

If the complex rank of $\Lambda \subset \mathbf{C}^n$ is n and the real rank $n+q$, then after a change of coordinates we can get $e_1, \cdots, e_n \in \Lambda$ and then $\Lambda = \mathbf{Z}^n \oplus \Gamma$ with a discrete subgroup $\Gamma \subset \mathbf{C}^n$ of real rank q. We say that $\Lambda = \mathbf{Z}^n \oplus \Gamma$ has the **rank** $n+q$.

Let $\pi : \mathbf{C}^n \to \mathbf{C}^n/\Lambda$ be the natural projection.

$$\begin{array}{ccl} \mathbf{C}_\Lambda & = & \mathbf{R}_\Lambda + i\mathbf{R}_\Lambda \\ | & & \\ \mathbf{R}_\Lambda & & \\ | & & \\ \mathbf{MC}_\Lambda & = & \mathbf{R}_\Lambda \cap i\mathbf{R}_\Lambda \end{array}$$

The maximal compact real subgroup of $\mathbf{C}^n/\Lambda$ is the **maximal real torus** $K := \pi(\mathbf{R}_\Lambda) := \mathbf{R}_\Lambda/\Lambda$, that is the projection of the real span

$$\begin{array}{c} K = \mathbf{R}_\Lambda/\Lambda \\ | \\ K_0 = \mathbf{MC}_\Lambda/(\mathbf{MC}_\Lambda \cap \Lambda) \end{array}$$

$\mathbf{R}_\Lambda$ of Λ. Moreover let $\mathbf{MC}_\Lambda := \mathbf{R}_\Lambda \cap i\mathbf{R}_\Lambda$ be the maximal $\mathbf{C}$–linear subspace of $\mathbf{R}_\Lambda$. Then $\mathbf{MC}_\Lambda \cap \Lambda$ is discrete in $\mathbf{MC}_\Lambda$ so that the projection $K_0 := \pi(\mathbf{MC}_\Lambda) = \mathbf{MC}_\Lambda/(\mathbf{MC}_\Lambda \cap \Lambda)$ becomes a complex subgroup of $\mathbf{C}^n/\Lambda$. K_0 is the **maximal complex subgroup** of the maximal real torus K.

1.1.3 Proposition
Let $\Lambda \subset \mathbf{C}^n$ be a discrete subgroup.

1. If the complex rank $m := \mathrm{rank}_{\mathbf{C}}\Lambda < n$ than
$$\mathbf{C}^n/\Lambda \simeq \mathbf{C}^{n-m} \oplus (\mathbf{C}^m/\Lambda)$$
where Λ is considered as a discrete subgroup of $\mathbf{C}^m$.
2. Let $\mathrm{rank}_{\mathbf{C}}\Lambda = n$ and $\mathrm{MC}_\Lambda = \mathbf{R}_\Lambda \cap i\mathbf{R}_\Lambda$ be the maximal complex subspace of $\mathbf{R}_\Lambda$. If the maximal complex subgroup $\mathrm{MC}_\Lambda/(\mathrm{MC}_\Lambda \cap \Lambda)$ is not dense in maximal real torus $\mathbf{R}_\Lambda/\Lambda$, then
$$\mathbf{C}^n/\Lambda \simeq \mathbf{C}^{*m} \oplus (\mathbf{C}^{n-m}/\Gamma)$$
when $\Gamma \subset \mathbf{C}^{n-m}$ is a discrete subgroup of complex rank $n-m$ and the maximal complex subgroup $\mathrm{MC}_\Gamma/(\mathrm{MC}_\Gamma \cap \Gamma)$ is dense in the maximal real torus $\mathbf{R}_\Gamma/\Gamma$ of $\mathbf{C}^{n-m}/\Gamma$.

Proof
1. If $m = \mathrm{rank}_{\mathbf{C}}\Lambda$, then Λ spans a complex subspace $V := \mathbf{C}_\Lambda \subset \mathbf{C}^n$ of complex dimension m so that $\mathbf{C}^n = U \oplus V$ with an $n-m$–dimensional $\mathbf{C}$–linear subspace U. We get $\mathbf{C}^n/\Lambda = U \oplus V/\Lambda$.
2. The closure $\overline{K_0} = \overline{\pi(\mathrm{MC}_\Lambda)} \subset \mathbf{R}_\Lambda/\Lambda$ of the projection of the maximal complex subspace $\mathrm{MC}_\Lambda = \mathbf{R}_\Lambda \cap i\mathbf{R}_\Lambda$ of $\mathbf{R}_\Lambda$ is a connected and closed real subgroup of the maximal real torus $K = \mathbf{R}_\Lambda/\Lambda$. So K splits into $K' = \overline{K_0}$ and another torus K'' of a certain real dimension m.
Because $K = K' \oplus K''$ we get the decomposition $\Lambda = \Lambda' \oplus \Lambda''$ such that $K' = \mathbf{R}_{\Lambda'}/\Lambda'$ and $K'' = \mathbf{R}_{\Lambda''}/\Lambda''$ where $\mathrm{rank}_{\mathbf{R}}\Lambda'' = m$. Since $\mathrm{MC}_\Lambda \subset \mathbf{R}_{\Lambda'}$, $\mathrm{MC}_{\Lambda'} = \mathrm{MC}_\Lambda$ and $\mathrm{MC}_{\Lambda''} = 0$. Then $\mathrm{rank}_{\mathbf{C}}\Lambda'' = m$.
We assumed $\mathrm{rank}_{\mathbf{C}}\Lambda = n$. So we get $\mathbf{C}^n = \mathbf{C}_{\Lambda'} \oplus \mathbf{C}_{\Lambda''}$, where the $\mathbf{C}$–spans $\mathbf{C}_{\Lambda'}, \mathbf{C}_{\Lambda''}$ have the complex dimensions $n-m$ or m, respectively.
The real and the complex rank of Λ'' coincide. Then $\mathbf{C}_{\Lambda''}/\Lambda'' \simeq \mathbf{C}^{*m}$. On the other hand, the projection of $\mathrm{MC}_{\Lambda'} = \mathrm{MC}_\Lambda$ is dense in the real torus $K' = \mathbf{R}_{\Lambda'}/\Lambda'$ which is maximal in $\mathbf{C}_{\Lambda'}/\Lambda'$. We get $\mathbf{C}_{\Lambda'}/\Lambda' \simeq \mathbf{C}^{n-m}/\Gamma$ with $\Gamma := \Lambda'$. This proves the proposition. *Q.E.D.*

A consequence of this proposition is that for every toroidal group $\mathbf{C}^n/\Lambda$ the lattice Λ has maximal complex rank n.

KOPFERMANN introduced in 1964 the concept of n–dimensional toroidal groups with the irrationality condition [64].

1.1.4 Theorem
Let $\Lambda \subset \mathbf{C}^n$ be a discrete subgroup of complex rank n and MC_Λ the maximal complex subspace of the real span $\mathbf{R}_\Lambda$. Then the following statements are equivalent:

1. $\mathbf{C}^n/\Lambda$ is toroidal.
2. There exists no $\sigma \in \mathbf{C}^n \setminus \{0\}$ so that the scalar product $\langle\sigma, \lambda\rangle \in \mathbf{Z}$ is integral for all $\lambda \in \Lambda$. **(Irrationality condition)**
3. The maximal complex subgroup $\mathbf{MC}_\Lambda/(\mathbf{MC}_\Lambda \cap \Lambda)$ is dense in the maximal real torus $\mathbf{R}_\Lambda/\Lambda$ of $\mathbf{C}^n/\Lambda$. **(Density condition)**

Proof
1≻2. If a complex vector $\sigma \neq 0$ exists so that $\langle\sigma, \lambda\rangle \in \mathbf{Z}$ $(\lambda \in \Lambda)$, then the exponential function $\mathbf{e}(\langle\sigma, z\rangle) = \exp(2\pi i\langle\sigma, z\rangle)$ $(z \in \mathbf{C}^n)$ is a non–constant Λ–periodic holomorphic function.
2≻3. If the projection of the maximal complex subspace $\mathbf{MC}_\Lambda \subset \mathbf{R}_\Lambda$ is not dense in $\mathbf{R}_\Lambda/\Lambda$, then at least one $\mathbf{C}^*$ splits (Proposition 1.1.3). After change of coordinates there exists a $\mathbf{R}$–independent set of generators of Λ so that the unit vector e_1 is one and the others are orthogonal. Then for $\sigma := e_1$ all scalar products $\langle\sigma, \lambda\rangle \in \mathbf{Z}$ $(\lambda \in \Lambda)$.
3≻1. Let f be holomorphic on $\mathbf{C}^n/\Lambda$. Then f is bounded on the compact real torus $\mathbf{R}_\Lambda/\Lambda$ and therefore $f \circ \pi$ constant on the maximal complex subspace $\mathbf{MC}_\Lambda \subset \mathbf{R}_\Lambda$. Now f must be constant an $\mathbf{R}_\Lambda/\Lambda$ by the density condition. Then the pullback $f \circ \pi$ is constant on $\mathbf{R}_\Lambda$ and because the complex rank of Λ is n the holomorphic function f must be constant on $\mathbf{C}^n$. *Q.E.D.*

A complex Lie group is a **Stein group**, if the underlying complex manifold is a Stein manifold. The products $\mathbf{C}^\ell \times \mathbf{C}^{*m}$ are Abelian Stein groups.

Every connected Abelian *real* group is isomorphic to $\mathbf{R}^l \times (\mathbf{R}/\mathbf{Z})^m$ where the second factor is a real torus. For connected and Abelian *complex* Lie groups we get the following decomposition proved by REMMERT [64, cf KOPFERMANN 1964] and by MORIMOTO in 1965 [74].

1.1.5 Decomposition of Abelian Lie groups (REMMERT-MORIMOTO)
Every connected Abelian complex Lie group is holomorphically isomorphic to a

$$\mathbf{C}^\ell \times \mathbf{C}^{*m} \times X_0$$

with a toroidal group X_0. The decomposition is unique.
Proof
Existence. Propositions 1.1.2 and 1.1.3 together with Theorem 1.1.4.
Uniqueness. Let $X_j := S_j \times T_j$ $(j = 1, 2)$ where S_j are connected Abelian Stein groups and T_j toroidal. If $\phi : X_1 \to X_2$ is an isomorphism, then obviously $\phi(T_1) \subset T_2$ and therefore $\phi(T_1) = T_2$, T_1 and T_2 are isomorphic and thus $S_1 = X_1/T_1$ and $S_2 = X_2/T_2$ are isomorphic. *Q.E.D.*

A consequence of the decomposition theorem and the density condition for toroidal groups is

1.1.6 Lemma

For any connected Abelian complex Lie group X the following statements are equivalent:

1. X is a Stein group.
2. X is isomorphic to $\mathbf{C}^\ell \times \mathbf{C}^{*m}$.
3. there exists no connected complex subgroup of positive dimension in the maximal compact real subgroup of X.

(Stein group criterion for Abelian Lie groups)

MATSUSHIMA and MORIMOTO proved in 1960 the following generalization of this lemma [70]

1.1.7 Theorem (MATSUSHIMA–MORIMOTO)

Let X be a connected complex Lie group. Then the following statements are equivalent:

1. X is a Stein group.
2. The connected component of the center of X is a Stein group.
3. X has no connected complex subgroup of positive dimension in any maximal compact real subgroup of X. **(Stein group criterion)**

For the *proof* of this theorem refer to the original paper.

With the previous Lemma the Stein groups are exactly the connected complex Lie groups where the connected center is isomorphic to a $\mathbf{C}^\ell \times \mathbf{C}^{*m}$.

Complex homomorphisms

Complex homomorphisms of connected and Abelian complex Lie groups can be described by Hurwitz relations. Instead of tangent spaces the description by universal covering spaces in the Abelian case will be prefered. Consider first the

1.1.8 Proposition

For any complex homomorphism $\tau : \mathbf{C}^n/\Lambda \to \mathbf{C}^{n'}/\Lambda'$ with discrete subgroups $\Lambda \subset \mathbf{C}^n$ and $\Lambda' \subset \mathbf{C}^{n'}$ there exists a unique $\mathbf{C}$–linear map $\hat{\tau} : \mathbf{C}^n \to \mathbf{C}^{n'}$ with the commutative diagram

$$\begin{array}{ccc} \mathbf{C}^n & \xrightarrow{\hat{\tau}} & \mathbf{C}^{n'} \\ \downarrow^{\pi} & & \downarrow^{\pi'} \\ \mathbf{C}^n/\Lambda & \xrightarrow{\tau} & \mathbf{C}^{n'}/\Lambda' \end{array}$$

where $\pi : \mathbf{C}^n \to \mathbf{C}^n/\Lambda$ and $\pi' : \mathbf{C}^{n'} \to \mathbf{C}^{n'}/\Lambda'$ are the natural projections.
$\hat{\tau}$ is called the **lift** of τ.
Conversely, a $\mathbf{C}$–linear map $\hat{\tau} : \mathbf{C}^n \to \mathbf{C}^{n'}$ with $\hat{\tau}(\Lambda) \subset \Lambda'$ induces a complex homomorphism $\tau : \mathbf{C}^n/\Lambda \to \mathbf{C}^{n'}/\Lambda'$ such that the diagram becomes commutative.

Proof
By the path lifting theorem there exists a unique continuous map $\hat{\tau} : \mathbf{C}^n \to \mathbf{C}^{n'}$ with $\hat{\tau}(0) = 0$ such that $\pi' \circ \hat{\tau} = \tau \circ \pi$. Hence $\hat{\tau}$ becomes a complex homomorphism and that is a $\mathbf{C}$–linear map. *Q.E.D.*

Let $X := \mathbf{C}^n/\Lambda$ and $X' := \mathbf{C}^{n'}/\Lambda'$ and $\tau : X \to X'$ be a complex homomorphism. Then:

τ is a covering map, iff its lift $\hat{\tau}$ is bijective. Then X is a *covering group* of X'.
τ is an isomorphism, iff $\hat{\tau}$ is bijective and $\hat{\tau}(\Lambda) = \Lambda'$.
X is a complex Lie subgroup of X',
iff there exists an (injective) immersion and homomorphism $\tau : X \to X'$, iff $\hat{\tau} : \mathbf{C}^n \to \mathbf{C}^{n'}$ is injective and $\hat{\tau}(\Lambda) = \Lambda' \cap \hat{\tau}(\mathbf{C}^n)$.
X is a closed complex Lie subgroup of X', iff there is an embedding and homomorphism $\tau : X \to X'$.

Now let $P := (\lambda_1, \cdots, \lambda_r)$ be a basis of $\Lambda \subset \mathbf{C}^n$ and $P' := (\lambda'_1, \cdots, \lambda'_{r'})$ be a basis of $\Lambda' \subset \mathbf{C}^{n'}$. Then $\hat{\tau} : \mathbf{C}^n \to \mathbf{C}^{n'}$ can be described by the matrix relation

$$CP = P'M', \qquad \textbf{(Hurwitzrelations)}$$

where $C \in \mathrm{M}(n', n; \mathbf{C})$ is defined by $\hat{\tau}(z) = Cz$ $(z \in \mathbf{C}^n)$ and $M' \in \mathrm{M}(r', r; \mathbf{Z})$ is an integral matrix. Then:

The complex homomorphism τ is a covering map, iff $C \in \mathrm{GL}(n, \mathbf{C})$ is regular.
τ is bijective, iff $C \in \mathrm{GL}(n, \mathbf{C})$ and $M' \in \mathrm{GL}(r, \mathbf{Z})$ so that

$$P' = CPM \quad \text{with} \quad M =: M'^{-1} \in \mathrm{GL}(r, \mathbf{Z}).$$

The group $X = \mathbf{C}^n/\Lambda$ is a complex Lie subgroup of $X' = \mathbf{C}^{n'}/\Lambda'$, iff the rank $C = n$ and $C\Lambda = \Lambda' \cap \hat{\tau}(\mathbf{C}^n)$.

Holomorphic maps $\tau : \mathbf{C}^n/\Lambda \to \mathbf{C}^{n'}/\Lambda'$ of toroidal groups are essentially complex homomorphisms.

1.1.9 Proposition
Let $\tau : \mathbf{C}^n/\Lambda \to \mathbf{C}^{n'}/\Lambda'$ be a holomorphic map with $\tau(1) = 1'$, where $\mathbf{C}^n/\Lambda$ is toroidal, $\mathbf{C}^{n'}/\Lambda'$ any complex Abelian Lie group and where $1, 1'$ are the units of $\mathbf{C}^n/\Lambda$, $\mathbf{C}^{n'}/\Lambda'$, respectively. Then τ is a complex homomorphism.

Proof
By path lifting theorem there exists a holomorphic map $\hat{\tau} : \mathbf{C}^n \to \mathbf{C}^{n'}$ with $\hat{\tau}(0) = 0$ so that $\tau \circ \pi = \pi' \circ \hat{\tau}$, where π, π' are the canonical projections. For any $\lambda \in \Lambda$ the difference $\hat{\tau}(z + \lambda) - \hat{\tau}(z)$ must be constant, namely $\hat{\tau}(\lambda) \in \Lambda'$. Let $\lambda' = \tau(\lambda)$. Then

$$\hat{\tau}_j(z+\lambda) = \hat{\tau}_j(z) + \lambda_j \qquad (z \in \mathbf{C}^n)$$

for the components $\hat{\tau}_j$ of $\hat{\tau}$ $(j = 1, \cdots, n)$. Now the partial derivatives $\partial_k \hat{\tau}_j$ are Λ–periodic and therefore constant since $\mathbf{C}^n/\Lambda$ is a toroidal group. Then $\hat{\tau}$ is a $\mathbf{C}$–linear map and τ a complex homomorphism. *Q.E.D.*

The following proposition describes the *Stein factorization* for toroidal groups.

1.1.10 Proposition

Let $X = \mathbf{C}^n/\Lambda$ be toroidal, $X' = \mathbf{C}^{n'}/\Lambda'$ any complex Abelian Lie group and $\tau : X \to X'$ a complex homomorphism. Then the image $\tau(X)$ is a toroidal group. The connected component $(\ker \tau)_o$ of the kernel of τ induces a factorization

$$\begin{array}{ccc} X & & \\ \downarrow & \searrow^{\tau} & \\ X/(\ker\tau)_o & \longrightarrow & X' \end{array}$$

Proof

Let $\hat{\tau} : \mathbf{C}^n \to \mathbf{C}^{n'}$ be the lift of τ. Then $\hat{\tau}$ is $\mathbf{C}$–linear, the image $V := \hat{\tau}(\mathbf{C}^n) \subset \mathbf{C}^{n'}$ a $\mathbf{C}$–linear subspace and $V \cap \Lambda'$ discrete in V. Therefore $\hat{\tau}(\Lambda) \subset V \cap \Lambda'$ is discrete. The map $\hat{\tau} : \mathbf{C}^n \to \mathbf{C}^{n'}$ induces a homomorphism

$$X \to V/(V \cap \hat{\tau}(\Lambda)) \to V/(V \cap \Lambda') \hookrightarrow X'$$

The image $\tau(X) = V/(V \cap \Lambda') \subset X'$ must be toroidal because X has non-constant holomorphic functions. Moreover the map

$$V/(V \cap \hat{\tau}(\Lambda)) \to V/(V \cap \Lambda')$$

is a covering map and $X/(\ker \tau)_o \simeq V/(V \cap \hat{\tau}(\Lambda))$. *Q.E.D.*

Toroidal coordinates and $\mathbf{C}^{*n-q}$–fibre bundles

Standard coordinates are used in torus theory whereas toroidal coordinates respect the maximal complex subspace $\mathbf{MC}_\Lambda$ of the $\mathbf{R}$–span $\mathbf{R}_\Lambda$ of the lattice $\Lambda \subset \mathbf{C}^n$.

1.1.11 Standard coordinates

Let $\Lambda \subset \mathbf{C}^n$ be a discrete subgroup of complex rank n and real rank $n+q$.

After a linear change of the coordinates we obtain $\Lambda = \mathbf{Z}^n \oplus \Gamma$ with a set of $\mathbf{R}$–independent $\mathbf{Z}$–generators $\gamma_1, \cdots, \gamma_q \in \Gamma$ of Γ. Then

$$P = (I_n, G) = \begin{pmatrix} I_q & 0 & \hat{T} \\ 0 & I_{n-q} & \tilde{T} \end{pmatrix}$$

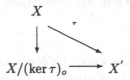

with unit $I_n := (e_1, \cdots, e_n) \in \mathrm{GL}(n, \mathbf{C})$ and $G := (\gamma_1, \cdots, \gamma_q) \in \mathrm{M}(n, q; \mathbf{C})$ are $\mathbf{R}$–independent, iff the rank of $\mathrm{Im}G$ is q. Thus, after a permutation of the coordinates we can assume that the square matrix $\hat{T}$ of the first q rows of G has an invertible imaginary part $\mathrm{Im}\hat{T}$. These coordinates are called **standard coordinates** of Λ. Of course they are not uniquely determined.

An immediate consequence of the irrationality condition 1.1.4(2) is:

A basis $P := (I_n, G)$ defines a *toroidal group*, iff the following condition holds:

(I) there exists no $\sigma \in \mathbf{Z}^n \setminus \{0\}$ so that ${}^t\sigma G \in \mathbf{Z}^n$.

(Irrationality condition in standard coordinates)

1.1.12 Toroidal coordinates

Toroidal coordinates where introduced by KOPFERMANN in 1964 and then by GHERARDELLI and ANDREOTTI in 1971/74. VOGT used them since 1981 and KAZAMA refined them in in 1984 slightly by transforming $\mathbf{MC}_\Lambda$ with $(\mathrm{Im}\hat{T})^{-1}$. [64, 33, 115, 116, 53]

Let $P = (I_n, G)$ be a standard basis of Λ so that the imaginary part of of the square matrix $\hat{T}$ of the first q rows of G is invertible, and let $\tilde{T}$ be the matrix of the last $n - q$ rows of G. Then change the first q coordinates u and the last $n - q$ ones v by the shear transformation

$$l(u, v) := (u, v + R_1 u) \qquad (u \in \mathbf{C}^q, v \in \mathbf{C}^{n-q})$$

where $R_1 := -(\mathrm{Im}\tilde{T})(\mathrm{Im}\hat{T})^{-1} \in \mathrm{M}(n - q, q; \mathbf{R})$. We get **toroidal coordinates**. After changing the order of the vectors the basis of the given lattice becomes

$$P = \begin{pmatrix} 0 & I_q & \hat{T} \\ I_{n-q} & R_1 & R_2 \end{pmatrix} = \begin{pmatrix} 0 & B \\ I_{n-q} & R \end{pmatrix}$$

where $B := (I_q, \hat{T}) \in \mathrm{M}(q, 2q; \mathbf{C})$ is the basis of a q–dimensional complex torus T. The real matrix $R := (R_1, R_2) \in \mathrm{M}(n - q, 2q; \mathbf{R})$ is the so–called **glueing matrix**.

The lattice becomes

$$\Lambda = \begin{pmatrix} 0 \\ \mathbf{Z}^{n-q} \end{pmatrix} \oplus \Gamma \text{withbasis} \begin{pmatrix} B \\ R \end{pmatrix} \text{of} \Gamma.$$

Toroidal coordinates have the following properties:

1. $\mathbf{MC}_\Lambda = \{z \in \mathbf{C}^n : \ z_{q+1} = \cdots = z_n = 0\}$ is the subspace of the first q coordinates,

2. $\mathbf{R}_\Lambda = \{z \in \mathbf{C}^n : \mathrm{Im} z_{q+1} = \cdots = \mathrm{Im} z_n = 0\} = \mathbf{MC}_\Lambda \oplus V$, where V is the real subspace generated by the $n-q$ units $e_{q+1}, \cdots, e_n \in \Lambda$,
3. $\mathbf{C}^n = \mathbf{MC}_\Lambda \oplus V \oplus iV = \mathbf{R}_\Lambda \oplus iV$.

Of course toroidal coordinates are not in the least uniquely determined, - toroidal groups have many symmetries.

We can transform the maximal complex subspace $\mathbf{MC}_\Lambda$ with $(Im\hat{T})^{-1}$. That is the same as if we transform the standard coordinates by

$$l(u,v) := ((\mathrm{Im}\hat{T})^{-1}u, v + R_1 u) \qquad (u \in \mathbf{C}^q, v \in \mathbf{C}^{n-q}).$$

After changing the order of the basis periods we obtain

$$P = \begin{pmatrix} 0 & B_1 \; B_2 \\ I_{n-q} & R_1 \; R_2 \end{pmatrix} = \begin{pmatrix} 0 & B \\ I_{n-q} & R \end{pmatrix},$$

where now $B_1 := (\mathrm{Im}\hat{T})^{-1}$ and $B_2 := (\mathrm{Im}\hat{T})^{-1}\mathrm{Re}\hat{T} + iI_q$. Then $B := (B_1, B_2)$ is a basis of the same torus T as before and $R := (R_1, R_2)$ the same glueing matrix. The advantage of these toroidal coordinates with **refined transformation** from standard coordinates is

$$(\dagger) \qquad l(\mathrm{Im}\gamma_j) = e_j \; (j = 1, \cdots, q) \mathrm{and} l(e_j) = e_j \; (j = q+1, \cdots, n)$$

for the basis $\gamma_1, \cdots, \gamma_q$ of Γ.

In toroidal coordinates the lattice Λ defines a *toroidal group*, iff the following condition on the glueing matrix R holds:

(I) there exists no $\sigma \in \mathbf{Z}^{n-q} \setminus \{0\}$ such that ${}^t\sigma R \in \mathbf{Z}^{2q}$.

(Irrationality condition in toroidal coordinates)

It is to remark that this condition depends only on the glueing matrix R and not on the torus T generated by B.

1.1.13 Real parametrizations

Toroidal coordinates allow a simple *real* parametrization of $\mathbf{C}^n/\Lambda$. For this let $\lambda_1, \cdots, \lambda_n$ be the first and $\gamma_1, \cdots, \gamma_q$ the last q elements of P so that we can complete the basis by $\gamma_j := ie_j$ $(j = q+1, \cdots, n)$ to a $\mathbf{R}$–basis of the $\mathbf{R}^{2n}$. Then the $\mathbf{R}$–linear map

$$(\hat{L}) \qquad z = \hat{L}(t) = \sum_{j=1}^{n} (\lambda_j t_j + \gamma_j t_{n+j}) \qquad (t \in \mathbf{R}^{2n})$$

induces a *real* Lie group isomorphism $L : \mathbf{T} \times \mathbf{R}^{n-q} \to \mathbf{C}^n/\Lambda$, where $\mathbf{T} := (\mathbf{R}/\mathbf{Z})^{n+q}$.

If we denote with u the first q toroidal coordinates and with v the last $n-q$ ones, then the real toroidal coordinates Reu, Imu, Rev, Imv are given after a change of the real parameters $t_1, \cdots, t_{2n}$ by

$$(\hat{L}_{\mathbf{R}}) \qquad \hat{L}_{\mathbf{R}}(t) = At \qquad (t \in \mathbf{R}^{2n})$$

with

$$A := \begin{pmatrix} I_q & \mathrm{Re}\hat{T} & 0 & 0 \\ 0 & \mathrm{Im}\hat{T} & 0 & 0 \\ R_1 & R_2 & I_{n-q} & 0 \\ 0 & 0 & 0 & I_{n-q} \end{pmatrix} \text{ or } \begin{pmatrix} (\mathrm{Im}\hat{T})^{-1} & (\mathrm{Im}\hat{T})^{-1}\mathrm{Re}\hat{T} & 0 & 0 \\ 0 & I_q & 0 & 0 \\ R_1 & R_2 & I_{n-q} & 0 \\ 0 & 0 & 0 & I_{n-q} \end{pmatrix}.$$

In the second case we get real toroidal coordinates as given by refined transformation from standard coordinates. In both cases functions $\mathbf{Z}^{n+q}$-periodic in the first $n+q$ real t-variables become Λ–periodic.

1.1.14 $\mathbf{C}^{*n-q}$–fibre bundles over a torus

Toroidal coordinates define a representation of any toroidal group $\mathbf{C}^n/\Lambda$ with rank $n+q$ as $\mathbf{C}^{*n-q}$*–fibre bundle.* The projection $\hat{p} : \mathbf{C}^n \to \mathrm{M}\mathbf{C}_\Lambda$ onto the subspace $\mathrm{M}\mathbf{C}_\Lambda$ of the first q variables induces a complex homomorphism

$$p : X = \mathbf{C}^n/\Lambda \to T = \mathrm{M}\mathbf{C}_\Lambda / B\mathbf{Z}^{2q}$$

onto the torus $T := \mathrm{M}\mathbf{C}_\Lambda/B\mathbf{Z}^{2q}$ with kernel $\mathbf{C}^{n-q}/\mathbf{Z}^{n-q} \simeq \mathbf{C}^{*n-q}$ closed in X so that the diagram

$$(\dagger) \qquad \begin{array}{ccc} \mathbf{C}^n & \xrightarrow{\hat{p}} & \mathrm{M}\mathbf{C}_\Lambda \\ \downarrow{\scriptstyle \pi'} & & \downarrow{\scriptstyle \pi} \\ X & \xrightarrow{p} & T \end{array}$$

becomes commutative.

It is well known that every closed complex Lie subgroup N of a Lie group X defines a principal fibre bundle with base space X/N and fibre N (see STEENROD [105, §7.4] or HIRZEBRUCH [45, §3.4]).

Thus, as an immediate consequence of toroidal coordinates every Lie group $X = \mathbf{C}^n/\Lambda$ with $\Lambda = \mathbf{Z}^n \oplus \Gamma$ of rank $n+q$ is a principal $\mathbf{C}^{*n-q}$–fibre bundle over the complex q–dimensional torus T as base space.

Such a bundle is defined by an *automorphic factor* α_τ $(\tau \in B\mathbf{Z}^{2q})$, fulfilling the *cocycle condition* $\alpha_{\tau+\tau'}(z) = \alpha_{\tau'}(z+\tau)\alpha_\tau(z)$ $(z \in \mathrm{M}\mathbf{C}_\Lambda, \tau, \tau' \in B\mathbf{Z}^{2q})$.

1.1.15 Lemma
Let $X = \mathbf{C}^n/\Lambda$ be of rank $n+q$ and let Λ in toroidal coordinates be $\mathbf{Z}$–generated by

$$P = \begin{pmatrix} 0 & B \\ I_{n-q} & R \end{pmatrix}$$

with the basis $B = (\tau_1, \cdots, \tau_{2q}) \in \mathrm{M}(q, 2q; \mathbf{C})$ of the torus group $T = \mathbf{MC}_\Lambda/B\mathbf{Z}^{2q}$ and the glueing matrix $R = (r_1, \cdots, r_{2q}) \in \mathrm{M}(n-q, 2q; \mathbf{R})$.

Then the principal $\mathbf{C}^{*n-q}$–fibre bundle $X \overset{p}{\to} T$ is given by the constant automorphic factor

$$(*) \qquad \alpha_{\tau_k} := \mathbf{e}(r_k) \in (S_1)^{n-q} \subset \mathbf{C}^{*n-q} \qquad (k = 1, \cdots, 2q),$$

where $\mathbf{e} : \mathbf{R}^{n-q} \to \mathbf{C}^{*n-q}$ is the exponential map and $S_1 := \{z \in \mathbf{C} : |z| = 1\}$ the 1–sphere. The bundle is topologically trivial.

Proof
Let $u \in \mathbf{MC}_\Lambda$ be the first q and $v \in \mathbf{C}^{n-q}$ the last $n-q$ variables of the $\mathbf{C}^n = \mathbf{MC}_\Lambda \oplus \mathbf{C}^{n-q}$. The pullback of the principal bundle $X \overset{p}{\to} T$ along the projection $\pi : \mathbf{MC}_\Lambda \to T = \mathbf{MC}_\Lambda/B\mathbf{Z}^{2q}$ is

$$\pi^* X = \{(u, x) \in \mathbf{MC}_\Lambda \times X : \pi(u) = p(x)\}.$$

Then we get a trivialization

$$\iota : \pi^* X \to \mathbf{MC}_\Lambda \times \mathbf{C}^{*n-q} \quad by \quad \iota(u, x) := (u, \mathbf{e}(v)) \quad (u \in \mathbf{MC}_\Lambda,\ x \in X),$$

where $v \in \mathbf{C}^{n-q}$ is any v with $\binom{u}{v} \in \pi'^{-1}(x)$ by (†).

Indeed, $v_1 \equiv v_2 \mod \mathbf{Z}^{n-q}$ iff $\binom{u}{v_1} \equiv \binom{u}{v_2} \mod \Lambda$ so that ι becomes a bundle isomorphism.

Now let $\tau = B\sigma$ a period with a $\sigma \in \mathbf{Z}^{2q}$. Define $\tau(u, x) := (u + \tau, x)$ for any $\binom{u}{v} \in \hat{\pi}'^{-1}(x)$. Instead of $\binom{u}{v}$ we can take a $\binom{u+B\sigma}{v+R\sigma} \in \hat{\pi}'^{-1}(x)$ so that

$$\iota \circ \tau(u, x) = (u + B\sigma, \mathbf{e}(v + R\sigma)) = (u + B\sigma, \mathbf{e}(R\sigma) \circ \mathbf{e}(v))$$

where $\mathbf{e}(R\sigma)$ acts componentwise. Therefore the constant automorphic factor $\alpha_{B\sigma} := \mathbf{e}(R\sigma)$ $(\sigma \in \mathbf{Z}^n)$ defines $X \overset{p}{\to} T$.

Let L_j be the topologically trivial line bundle $L_j \overset{p_j}{\to} T$ defined by the automorphic factor

$$(**) \qquad \alpha_j(\tau_k) = \mathbf{e}(r_{jk}) \qquad (k = 1, \cdots, 2q)$$

where $R = (r_{jk}) \in \mathrm{M}(n-q, 2q; \mathbf{R})$ is the given glueing matrix. Then the topologically trivial sum $L = L_1 \oplus \cdots \oplus L_{n-q}$ is a vector bundle $L \overset{\oplus p_i}{\to} T$, defined by the automorphic factor $\alpha = \operatorname{diag}(\alpha_1, \cdots, \alpha_{n-q}) \in \mathrm{GL}(n-q; \mathbf{C})$ and associated to the given bundle $X \xrightarrow{p} T$ which is then topologically trivial. *Q.E.D.*

If the line bundles L_α, L_β are defined by $\alpha_\lambda, \beta_\lambda$, respectively, then the ***tensor product*** $L_\alpha \otimes L_\beta$ is defined by the product $\alpha_\lambda \beta_\lambda$ $(\lambda \in \Lambda)$. If a line bundle L is defined by an automorphic factor α_λ, then the ***dual bundle*** L^* is defined by α_λ^{-1} $(\lambda \in \Lambda)$. If $\tau : \mathbf{C}^n/\Lambda \to \mathbf{C}^{n'}/\Lambda'$ is a complex homomorphism and the line bundle L' on $\mathbf{C}^{n'}/\Lambda'$ defined by the autmorphic factor $\alpha'_{\lambda'}$, then the ***pullback*** $\tau^* L'$ on $\mathbf{C}^n/\Lambda$ is defined by the automorphic factor $\alpha_\lambda := \hat{\tau}^* \alpha'_{\hat{\tau}(\lambda)}$.

1.1.16 Remark

Let $X = \mathbf{C}^n/\Lambda$ as in the previous lemma be represented as natural $\mathbf{C}^{*n-q}$–fibre bundle over a torus T and L_j as in the previous proof be the line bundle $L_j \xrightarrow{p_j} T$ defined by the automorphic factor

$$\alpha_j(\tau_k) = \mathbf{e}(r_{jk}) \qquad (k = 1, \cdots, 2q)$$

where $R = (r_{jk}) \in \mathrm{M}(n-q, 2q; \mathbf{R})$ is the given glueing matrix.
Then we get by the irrationality condition (I) for toroidal groups:

X is a toroidal group, iff for all $\sigma \in \mathbf{Z}^{n-q} \setminus \{0\}$ the topologically trivial line bundle

$$L^\sigma := \bigotimes_{j=1}^{n-q} L_j^{\otimes \sigma_j} \quad \text{givenby} \quad \alpha^\sigma(\tau_k) = \mathbf{e}(\sum_{j=1}^{n-q} \sigma_j r_{jk}) \qquad (k = 1, \cdots, 2q)$$

is not analytically trivial. **(Irrationality condition for line bundles)**

Maximal Stein subgroups of toroidal groups

We have seen that every group $X = \mathbf{C}^n/\Lambda$ with $\Lambda = \mathbf{Z}^n \oplus \Gamma$ of rank $n+q$ has $\mathbf{C}^{*n-q}$ as closed subgroup. We shall see that the dimension $n-q$ is maximal for toroidal groups.

Serre conjectured in 1953 that a complex analytic fibre space whose base and fibre are Stein manifolds is a Stein manifold [100]. Matsushima and Morimoto proved in 1960 that a complex analytic fibre bundle X is a Stein manifold, if base space B and fibre F are Stein manifolds and its structure group G is a connected complex Lie group. A principal bundle is a Stein manifold, if base space B and structure group G are connected complex Stein groups. [70]

1.1.17 Proposition
Let $X = \mathbf{C}^n/\Lambda$ be a toroidal group of rank $n+q$. Then:

1. For every closed Stein subgroup $N \doteq \mathbf{C}^\ell \times \mathbf{C}^{*m}$ we have $2\ell + m \leq n - q$.
2. If $N \subset X$ is a maximal closed Stein subgroup, then X/N is a torus group.

Proof
1. Let $V = \mathbf{C}^{\ell+m}$ be the universal covering of N so that the inclusion $N \hookrightarrow X$ induces the inclusion of the lift $V \hookrightarrow \mathbf{C}^n$. Then $V \cap \Lambda$ has rank m because $N = V/(V \cap \Lambda)$ has the same rank. So $(V + \Lambda)/V$ has the rank $n + q - m$. Now the quotient X/N must be a toroidal group because X is toroidal.
Because $X/N = (\mathbf{C}^n/V)/((V + \Lambda)/V)$ the rank of X/N must be $n + q - m$. On the other hand the dimension of X/N is $n - \ell - m$ and therefore its rank $\leq 2(n - \ell - m)$. So $n + q - m \leq 2(n - \ell - m)$ what was to be proved.
2. If X/N is a non-compact toroidal group it contains a closed Stein subgroup $S \simeq \mathbf{C}^{*r}$ with $r > 0$ as we have seen in the section before. Then $\pi : X \to X/N$ induces a principal N-fibre bundle $\pi^{-1}(S)$ with Stein manifold S as base space. With the mentioned result of Matsushima and Morimoto $\pi^{-1}(S)$ is a Stein submanifold bigger than N so that N cannot be maximal. X/N must be a compact torus group. *Q.E.D.*

We shall see later (p 70) that quasi-Abelian varieties can have maximal closed Stein subgroups isomorphic to $\mathbf{C}^\ell \times \mathbf{C}^{*m}$ with $2\ell + m = n - q$ and $\ell > 0$.

1.2 Toroidal subgroups and pseudoconvexity

Every complex Lie group has a maximal toroidal subgroup which determines the type q of the group. Such a group is exactly $(n-q)$–complete. Some toroidal groups have a maximal torus subgroup. Compact analytic sets of toroidal groups are contained in translations of these torus subgroups.

The maximal toroidal subgroup of a complex Lie group

Let X be an n–dimensional connected complex Lie group. Then $p \in X$ is a **left period** of a meromorphic function f on X if

$$f(px) = f(x) \qquad \text{forall} x \in X.$$

The **period group** P_f is the closed complex subgroup of all left periods of f and the **rank** of f is

$$\text{rank} f := n - \dim P_f.$$

The function f is **non–degenerate**, if rank$f = n$ otherwise **degenerate** and **totally degenerate**, if f is constant.
The **period group** of all meromorphic or holomorphic functions is the intersection of the period groups of all functions meromorphic or holomorphic, respectively, on X.

The *type* is an important invariant of complex Lie groups. First, we give the definition of the type for toroidal groups, which also holds for any connected Abelian complex Lie group.

1.2.1 Definition
A toroidal group $X = \mathbf{C}^n/\Lambda$ is of **type** q, if the maximal complex subspace $\mathbf{MC}_\Lambda$ of the real span $\mathbf{R}_\Lambda$ has the complex dimension q.

For a toroidal group $X = \mathbf{C}^n/\Lambda$ with $\Lambda = \mathbf{Z}^n \oplus \Gamma$ the type is the real rank q of the lattice Γ.

It is well known that a *complex Lie group* X is a Stein group under *one* of the conditions:

a) X is holomorphically separable.
b) X has at every point local coordinates given by global functions.

MORIMOTO [74] proved in 1965:

1.2.2 Theorem (Holomorphic reduction) (MORIMOTO)
Let X be a connected complex Lie group with unit 1, and let

$$X_0 := \{a \in X : \ f(a) = f(1) \text{ for all holomorphic functions f in } X\}$$

be the constant set of all holomorphic functions. Then:

1. X_0 is the period group of all holomorphic functions.
2. X_0 is the smallest closed normal complex subgroup such that X/X_0 is a Stein group.
3. X_0 is a toroidal subgroup contained in the center of X.
4. Every complex homomorphism $\phi : X \to Y$ into a Stein group Y can be split as $\phi = \psi \circ \pi$ with the natural projection $\pi : X \to X/X_0$ and a complex homomorphism $\psi : X/X_0 \to Y$ of Stein groups.

X_0 is called the **maximal toroidal subgroup** of X and
X/X_0 the **holomorphic reduction** of X.
X is said to be of **type** q, if its maximal toroidal subgroup X_0 is of type q.
Proof
a) X_0 is *closed* in X: X_0 is the intersection of the period groups of all holomorphic functions.
X_0 is a *subgroup* of X: For $a \in X$ and holomorphic f the functions $f_a(x) := f(ax)$ and $f_-(x) := f(x^{-1})$ $(x \in X)$ are holomorphic. For $a, b \in X_0$ we get $f(ab) = f_a(b) = f_a(1) = f(a) = f(1)$ and $f(a^{-1}) = f_-(a) = f_-(1) = f(1)$ such that $ab, a^{-1} \in X_0$.
X_0 is a *complex subgroup* of X: Let $\mathcal{X}_0$ and $\mathcal{X}$ be the Lie algebras of X_0, X, respectively. We want to prove that $\mathcal{X}_0$ is a *complex* subalgebra of $\mathcal{X}$. Let $U \in \mathcal{X}_0$ and $V := iU \in \mathcal{X}$. Moreover let $\exp sU$ be the map which maps $s \in \mathbf{R}$ on a 1–parameter subgroup of X with tangent U at 1. Then define

$$\psi(s + it) := \exp(sU)\exp(tV)$$

in a certain connected neighborhood of $0 \in \mathbf{C}$.
For a holomorphic function $f : X \to \mathbf{C}$ let $f_0 := f \circ \psi$. Now $\exp sU \in X_0$ so that $f_0(s) = f_0(0)$ and then $f_0(z) = f_0(0)$ locally around $0 \in \mathbf{C}$. Then $\exp tV \in \mathcal{X}_0$ for all $t \in \mathbf{R}$ so that $V \in \mathcal{X}_0$.
X_0 is a *characteristic subgroup* of X: For $\sigma \in \mathrm{Aut}X$ and a holomorphic function f we define the holomorphic function $f^\sigma(x) := f(\sigma(x))$ $(x \in X)$. For $a \in X_0$ we get $f(\sigma(a)) = f^\sigma(a) = f^\sigma(1) = f(1)$ so that $\sigma(a) \in X_0$.
X_0 is the *period group of all holomorphic functions*: Suppose $f(px) = f(x)$ $(x \in X)$ for a fixed $p \in X$ and all holomorphic functions f. Then $f(p) = f(1)$ so that $p \in X_0$. On the other hand let $f(p) = f(1)$ for all f. Then $f(px) = f(xx^{-1}px) = f_x(x^{-1}px) = f_x(1) = f(1)$ $(x \in X)$, hence p is a period for all holomorphic functions.

b) X/X_0 is a *Stein group*: As we mentioned before it is sufficient to show that X/X_0 is holomorphically separable. By the definition of X_0 we can define the

natural homomorphism $\varphi : \mathcal{H}(X) \to \mathcal{H}(X/X_0)$ with $\varphi(f) \circ \pi = f$ for $f \in \mathcal{H}(X)$, where $\pi : X \to X/X_0$ is the projection. Let $\overline{a} = \pi(a)$ and $\overline{b} = \pi(b)$ be distinct elements of X/X_0. Then there exists $f \in \mathcal{H}(X)$ such that $f(a^{-1}b) \neq f(1)$ for $a^{-1}b \notin X_0$. We set $\overline{f} := \varphi(f_{a^{-1}}) \in \mathcal{H}(X/X_0)$. The $\overline{f}$ separates $\overline{a}$ and $\overline{b}$.
X_0 is the *smallest closed and normal complex subgroup* of X, such that X/X_0 is a Stein group: Let N be a closed and normal subgroup of X such that X/N is a Stein group. Take $a \in X \setminus N$. Moreover let $f^* \in \mathcal{H}(X/N)$ such that $f^*(\pi(a)) \neq f^*(\pi(1))$ with natural projection $\pi : X \to X/N$. Then $f := f^* \circ \pi$ separates a and 1. Hence $a \notin X_0$, so $X_0 \subset N$.
X_0 is *connected*: Let X_0^0 be the connected component of X_0 containing the unit 1. Then X_0^0 is normal in X and X/X_0^0 a covering group of X/X_0. By a result of Stein, X/X_0^0 is a Stein group. Then $X_0 \subset X_0^0$, therefore $X_0 = X_0^0$ is connected.
On X_0 *all holomorphic functions are constant*: Let $X_{00} \subset X_0$ be the constant set of 1 of all functions holomorphic in X_0. Then X_{00} is closed and normal in X_0. So X/X_0 and X_0/X_{00} are Stein groups and $X/X_0 \simeq (X/X_{00})/(X_0/X_{00})$. X/X_{00} can be considered as a principle fibre bundle whose base space and fibre are Stein groups. So by the mentioned result of MATSUSHIMA and MORIMOTO on p 14 X/X_{00} is a Stein group. Then $X_0 \subset X_{00}$, hence $X_{00} = X_0$. All holomorphic functions on X_0 are constant.
X_0 is *in the center* Z of X: Let $\mathcal{X}$ be the Lie algebra of X. The image of the adjoint representation $\mathrm{Ad} : X \to \mathrm{GL}(\mathcal{X})$ of X is a Lie subgroup of $\mathrm{GL}(\mathcal{X})$ and the kernel is the center Z of X. It is well known that $\mathrm{GL}(\mathcal{X})$ is a Stein group. Thus X/Z is Stein. By the previous result X_0 is the smallest closed and normal subgroup such that X/X_0 is a Stein group. Then $X_0 \subset Z$.

c) Let $\phi : X \to Y$ be a homomorphism of a complex Lie group X into a Stein group Y and $X_0 \subset X$ be the maximal toroidal subgroup of X. Then $\phi(X_0)$ is a connected and Abelian complex subgroup of Y and then toroidal so that $X_0 \subset \ker\phi$. But then $\phi = \psi \circ \pi$ with $\psi : X/X_0 \to Y$. *Q.E.D.*

1.2.3 Remark

Every complex Lie group with only constant holomorphic functions is Abelian and connected and therefore a toroidal group $\mathbf{C}^n/\Lambda$.
Every compact and connected complex Lie group is a torus.

The holomorphic function rings of complex Lie groups are those of Stein groups.

For a connected complex Lie group X with Lie Algebra $\mathcal{X}$ let K be a maximal real compact subgroup with Lie algebra $\mathcal{K}$. Moreover let $\mathcal{K}_0$ be the maximal complex subalgebra of $\mathcal{K}$. Then there exists a uniquely defined complex subgroup K_0 of K associated with $\mathcal{K}_0$ which is independent of the choice of K [75]. We get the following diagram in which we note first the general and behind the Abelian case. The lattice Λ in the diagram gives always a toroidal group.

	Lie group		Lie algebra	
	general	Abelian	general	Abelian
group of dim n	X	$\mathbf{C}^{\ell} \times \mathbf{C}^{*m} \times X_0$	$\mathcal{X}$	$\mathbf{C}^{\ell+m+n_0}$
|	|	|	|	|
max. toroidal subgroup of dim n_0	X_0	$\mathbf{C}^{n_0}/\Lambda = \mathbf{C}_\Lambda/\Lambda$	$\mathcal{X}_0 = \mathcal{K} + i\mathcal{K}$	$\mathbf{C}^{n_0} = \mathbf{R}_\Lambda + i\mathbf{R}_\Lambda$
|	|	|	|	|
max. compact real subgroup of real dim n_0+q	K	$\mathbf{R}_\Lambda/\Lambda$	$\mathcal{K}$	$\mathbf{R}_\Lambda$
|	|	|	|	|
max. complex subgroup of dim q	K_0	$\mathrm{M}\mathbf{C}_\Lambda/(\mathrm{M}\mathbf{C}_\Lambda \cap \Lambda)$	$\mathcal{K}_0 = \mathcal{K} \cap i\mathcal{K}$	$\mathrm{M}\mathbf{C}_\Lambda = \mathbf{R}_\Lambda \cap i\mathbf{R}_\Lambda$

The **maximal toroidal subgroup** X_0 and the **type** q **of a Lie group** X

Another immediate consequence of the previous theorem is the

1.2.4 Lemma

Let X be a connected complex Lie group with the maximal toroidal subgroup X_0. Then X is holomorphically convex, iff X_0 is compact.

Proof

$\succ$. If X_0 is not compact let $a \in X_0$. Then the holomorphically convex hull of $\{a\}$ is X_0. So X is not holomorphically convex.

$\prec$. X_0 is a closed subgroup of X so that X/X_0 is a Stein group. Now X is a principal bundle with base space X/X_0 and fibre X_0. Then X is holomorphically convex. *Q.E.D.*

Pseudoconvexity

Let X be a complex manifold of complex dimension n.

A real $\mathcal{C}^2$–function $\phi : X \to \mathbf{R}$ is **plurisubharmonic**, iff the complex Hessian form

$$H(\phi)(x) := \sum_{j,k=1}^{n} \frac{\partial^2 \phi}{\partial z_j \partial \overline{z}_k}(x) \zeta^j d\overline{\zeta^k} \qquad \textbf{(Levi form)}$$

is positive semi–definite for all $x \in X$, p–**plurisubharmonic** iff ϕ has in addition at least p positive eigenvalues for all $x \in X$ and **strictly plurisubharmonic**, iff it is n-plurisubharmonic.

The $\mathcal{C}^2$–function $\phi : X \to \mathbf{R}$ is an **exhausting** function for X, iff

$$\{x \in X : \ \phi(x) < c\} \subset\subset X \qquad \text{for all real } c.$$

These p–plurisubharmonic functions are special cases of p–*convex* functions which have at least p positive eigenvalues, but that are not necessairely positive semi–definite. It is to remark that in literature the fact that a function is p–convex is counted in different ways. So GRAUERT and ANDREOTTI counted $n-p+1$ eigenvalues for the benefit of the formulation of a cohomology theorem. We follow HENKIN and LEITERER with p eigenvalues so that ϕ becomes linearly convex in the first p of suitable coordinates [44, §4.3].

Plurisubharmonic functions obey the maximum principle for connected analytic sets. The exhaustion property for plurisubharmonic functions implies that every compact set $K \subset X$ has a compact plurisubharmonic hull.

1.2.5 Definition
A complex manifold X is **p–complete**, iff it has an exhausting p–plurisubharmonic function. X is **purely p–complete**, iff it is p–complete but not $(p+1)$–complete.

These manifolds are special *completely p–convex* manifolds. Those are defined by exhausting p–convex functions only. GRAUERT showed in 1958 with a supplement of NARASIMHAN in 1961 that Stein manifolds are n–complete. Compact manifolds are 0–complete but never 1–complete because all plurisubharmonic functions are constant.

It is to remark, that here also the counting differs in literature. So completely p–convex means also, that X can be exhausted by a $(p+1)$–convex function [44, §5.1]. Therefore a complex manifold is said to be **weakly 1–complete**, if it has an exhausting plurisubharmonic $\mathcal{C}^\infty$-function only.

For toroidal groups of type q, the real dimension $n-q$ of the space outside the maximal real subtorus $\mathbf{R}_\Lambda/\Lambda$ of real dimension $n+q$ is important. KAZAMA proved in 1973 [52]:

1.2.6 Proposition
Every Abelian Lie group $X = \mathbf{C}^n/\Lambda$ of type q is purely $(n-q)$–complete.
Proof
In toroidal coordinates the $\mathbf{R}$–span of Λ is

$$\mathbf{R}_\Lambda = \{z = x + iy \in \mathbf{C}^n : \ y_{q+1} = \cdots = y_n = 0\}.$$

The Λ–periodic plurisubharmonic $\mathcal{C}^\infty$–function

$$\hat{\phi}(z) := \sum_{k=q+1}^{n} y_k^2 \qquad (z = x + iy \in \mathbf{C}^n)$$

induces a $\mathcal{C}^\infty$–function $\phi \geq 0$ on $X = \mathbf{C}^n/\Lambda$, which is exhausting on the group $X = \mathbf{C}^n/\Lambda$, because $\left\{z \in \mathbf{C}^n : \ \hat{\phi}(z) < c^2\right\} \subset \mathbf{R}_\Lambda \times D_c$ with $D_c :=$

$\{|y_k| \le c,\ k = q+1, \cdots, n\}$ for every real $c \ge 0$. Moreover ϕ is $(n-q)$–plurisubharmonic because the Levi form

$$\sum_{j,k=1}^{n} \frac{\partial^2 \hat{\phi}}{\partial z_j \partial \overline{z}_k} \zeta^j \overline{\zeta^k} = \frac{1}{4} \sum_{j,k=1}^{n} \frac{\partial^2 \hat{\phi}}{\partial y_j \partial y_k} \zeta^j \overline{\zeta^k} = \frac{1}{2} \sum_{k=q+1}^{n} \zeta^k \overline{\zeta^k}$$

is positive semi–definite with $n-q$ positive eigenvalues. On the other hand any plurisubharmonic function must be constant on $\mathbf{R}_\Lambda/\Lambda$ so that the rank of its lift cannot be bigger. $X = \mathbf{C}^n/\Lambda$ is purely $(n-q)$–complete. *Q.E.D.*

Consequently all non–compact toroidal groups are weakly 1–complete.

TAKEUCHI considered in 1974 all complex Lie groups [112].

1.2.7 Theorem (KAZAMA-TAKEUCHI)
Any connected complex Lie group X of complex dimension n and of type q is purely $(n-q)$–complete.

KAZAMA gave the following *proof* which we want to sketch only. As we have seen in the previous proposition there exists an $(n-q)$–plurisubharmonic exhaustion function ϕ on the maximal toroidal subgroup X_0 of X, constant on the maximal compact real subgroup of X_0. On the other hand X/X_0 is a Stein group (Theorem 1.2.2(2)) so that after the mentioned theorem of GRAUERT there exists an n–plurisubharmonic exhaustion function ψ on X/X_0. Then it is possible to combine both for a $(n-q)$–plurisubharmonic exhausting function on X.

The following corollary was proved by KAZAMA in 1971 [50]. For that we remember lemma 1.2.4 that X is holomorphically convex, iff the maximal toroidal subgroup X_0 is compact.

1.2.8 Corollary
Let X be a holomorphically convex complex Lie group of dimension n with compact maximal toroidal subgroup X_0 of dimension q. Then X is purely $(n-q)$–complete.

This is easy to explain. Because X/X_0 is a Stein group we have an $(n-q)$–plurisubharmonic exhausting function φ on X/X_0. Then its lift ϕ is $(n-q)$–plurisubharmonic on X. Because X is a principal bundle over X/X_0 with compact fibre, ϕ is exhausting.

The maximal complex subtorus of a complex Lie group

Every toroidal group has a uniquely defined maximal closed subtorus as Lie subgroup. But this torus may reduce to a point as MORIMOTO proved in [75].

1.2.9 Theorem (MORIMOTO)
Let X be a connected complex Lie group. Then there exists the unique maximal complex subtorus T as complex Lie subgroup, situated in $\pi(\mathbf{MC}_\Lambda) = \mathbf{MC}_\Lambda/(\mathbf{MC}_\Lambda \cap \Lambda)$, which contains all compact and connected complex analytic sets $A \subset X$ with $1 \in A$.

Proof
Of course every A must be part of the maximal toroidal subgroup X_0 of X, but we want to prove the theorem for all Abelian complex Lie groups $X = \mathbf{C}^n/\Lambda$. If $n = 1$, then $X = \mathbf{C}$, $\mathbf{C}^*$ or a torus T and the theorem is true. So let the theorem be true for all connected Abelian complex Lie groups of dimension $< n$ and $X = \mathbf{C}^n/\Lambda$ a group, defined by a discrete subgroup Λ of rank r. The cases $r = 0, 1$ and $2n$ are trivial so that we can assume $1 < r \leq 2n - 1$.
By the decomposition theorem 1.1.5 and toroidal coordinates 1.1.12 we get

$$\mathbf{C}^n = \mathbf{MC}_\Lambda \oplus (V' \oplus iV') \oplus W$$

where the maximal linear subspace $\mathbf{MC}_\Lambda$ of $\mathbf{R}_\Lambda$ is the subspace of the first q coordinates. $V = V' \oplus iV'$ is the linear subspace of the next $p := r - q$ coordinates, V' the $\mathbf{R}$–span of its unit vectors so that $\mathbf{R}_\Lambda = \mathbf{MC}_\Lambda \oplus V'$ and W the linear subspace of the last m coordinates.
Now it is easy to see that a connected compact analytic set $A \subset X$ with $1 \in A$ is in $\mathbf{R}_\Lambda/\Lambda$:
Let $z = u + w$ where $u \in \mathbf{MC}_\Lambda \oplus V' = \mathbf{R}_\Lambda$ and $w = \sum_{j=1}^{p+2m} x_j w_j \in iV' \oplus W$ with a $\mathbf{R}$–basis $w_1, \cdots, w_{p+2m}$. The functions

$$f_j : X \to \mathbf{R}, \text{ defined by } f_j(\pi(z)) = x_j \qquad (j = 1, \cdots, p + 2m)$$

are plurisubharmonic on X and therefore constant on A. So $A \subset \pi(\mathbf{R}_\Lambda) = \mathbf{R}_\Lambda/\Lambda$. Now we want to prove $A \subset \pi(\mathbf{MC}_\Lambda)$. Because $\pi(\mathbf{MC}_\Lambda) = \mathbf{MC}_\Lambda/(\mathbf{MC}_\Lambda \cap \Lambda)$ is an Abelian Lie group of dimension $< n$ we are ready.
So let $\hat{A}_0 \subset \mathbf{C}^n$ be the connected component of the lift $\hat{A} = \pi^{-1}(A)$ with $0 \in \hat{A}_0$. Now we consider $z = u + v$ with $u \in \mathbf{MC}_\Lambda$ and $v = \sum_{j=1}^{p} z_j v_j$ with a $\mathbf{C}$–basis $v_1, \cdots, v_p$ of $V = V' + iV'$. The functions

$$g_j(z) = v_j \qquad (j = 1, \cdots, p)$$

are real on $\hat{A}_0$ and therefore constant so that $\hat{A}_0 \subset \mathrm{MC}_\Lambda$ and $A = \pi(\hat{A}_0) \subset \pi(\mathrm{MC}_\Lambda)$. *Q.E.D.*

Every connected complex Lie group X has the unique maximal complex torus subgroup T which is contained *in the maximal complex subgroup of the maximal compact real subgroup of the maximal toroidal subgroup of* X. This maximal torus contains all compact and connected analytic sets $A \subset X$ with $1 \in A$.

Furthermore we get the following proposition as a consequence of the above argument.

1.2.10 Corollary
Let $X = \mathbf{C}^n/\Lambda$ be a toroidal group.
X has the positive dimensional maximal complex torus subgroup, iff $\pi(\mathrm{MC}_\Lambda) = \mathrm{MC}_\Lambda/(\mathrm{MC}_\Lambda \cap \Lambda)$ has a positive dimensional complex torus subgroup.

1.2.11 Definition
A complex Lie group without a torus subgroup of positive dimension is called **torusless.**

The maximal complex subgroup $\pi(\mathrm{MC}_\Lambda)$ of the maximal real compact subgroup $\mathbf{R}_\Lambda/\Lambda$ of a non–compact toroidal group $X = \mathbf{C}^n/\Lambda$ of type q has complex dimension $q < n$. Because of the density condition the maximal complex subgroup cannot be a torus. Therefore the maximal torus subgroup T has dimension $< q$, if X is not compact.

As an immediate consequence we get the

Example: Every toroidal group of dimension $n \geq 2$ and type 1 is torusless.

1.2.12 Proposition
For every toroidal group $X = \mathbf{C}^n/\Lambda$ with maximal subtorus T the quotient X/T is torusless.
Proof
Let $S \subset X/T$ be a positive dimensional subtorus. Then $\sigma^{-1}(S)$ is a compact complex submanifold of X, where $\sigma : X \to X/T$ is the projection. This is a contradiction for dim $\sigma^{-1}(S) >$ dim T. *Q.E.D.*

2. Line Bundles and Cohomology

Line bundles and cohomology of toroidal groups differ in a significant way from the compact torus case. KOPFERMANN *in 1964 used the characteristic decomposition of automorphic factors.* VOGT *studied in 1981/82 line bundles on toroidal groups in general. He distinguished between toroidal theta and wild groups.* VOGT *and* KAZAMA *calculated the cohomology groups of theta groups.* KAZAMA *and* UMENO *determined the cohomology of toroidal wild groups and even any complex Lie group by studying their maximal complex subgroups.*

2.1 Line bundles on toroidal groups

The characteristic decomposition of the exponential system which defines the automorphic factor of the line bundle is the tool with which we get the decomposition of the line bundle into a theta bundle and a topologically trivial line bundle. Toroidal groups can have topologically trivial line bundles which are not homogeneous.

Automorphic factors

Let $\Lambda \subset \mathbf{C}^n$ be a discrete subgroup.
An **automorphic factor** $\alpha : \Lambda \times \mathbf{C}^n :\longrightarrow \mathbf{C}^*$ or 1–cocycle of Λ is given by a set of holomorphic non–vanishing functions

$$\alpha_\lambda \in \mathcal{H}^*(\mathbf{C}^n) \qquad (\lambda \in \Lambda)$$

which fulfils the **cocycle condition**

$$\alpha_{\lambda+\lambda'}(z) = \alpha_{\lambda'}(z+\lambda)\alpha_\lambda(z) \qquad (z \in \mathbf{C}^n,\ \lambda, \lambda' \in \Lambda).$$

Under multiplication the automorphic factors form the Abelian Lie group of cocycles $Z^1(\Lambda, \mathcal{H}^*(\mathbf{C}^n))$. The automorphic factors $\alpha_\lambda, \tilde{\alpha}_\lambda$ $(\lambda \in \Lambda)$ are **cobordant,** if there exists a function $\psi \in \mathcal{H}^*(\mathbf{C}^n)$ with

$$\tilde{\alpha}_\lambda(z) = \psi(z+\lambda)\alpha_\lambda(z)\psi^{-1}(z) \qquad (z \in \mathbf{C}^n).$$

The automorphic factors cobordant to 1 form the subgroup of coboundaries $B^1(\Lambda, \mathcal{H}^*(\mathbf{C}^n))$ so that

$$H^1(\Lambda, \mathcal{H}^*(\mathbf{C}^n)) := Z^1(\Lambda, \mathcal{H}^*(\mathbf{C}^n))/B^1(\Lambda, \mathcal{H}^*(\mathbf{C}^n))$$

becomes the first cohomology group of Λ with values in $\mathcal{H}^*(\mathbf{C}^n)$.

Let X be a complex manifold and $\mathcal{O}$ the sheaf of germs of holomorphic functions. The *exponential sheaf sequence*

$$0 \to \mathbf{Z} \to \mathcal{O} \xrightarrow{\mathbf{e}} \mathcal{O}^* \to 0$$

with $\mathbf{e}(z) := \exp(2\pi i z)$ induces the long exact cohomology sequence

$$\cdots \to H^1(X, \mathcal{O}) \to H^1(X, \mathcal{O}^*) \xrightarrow{c_1} H^2(X, \mathbf{Z}) \to \cdots .$$

The **Picard group**

$$\mathrm{Pic}(X) := H^1(X, \mathcal{O}^*)$$

is the group of all isomorphism classes of holomorphic line bundles on X, the combining homomorphism

$$c_1 : H^1(X, \mathcal{O}^*) \to H^2(X, \mathbf{Z})$$

is the **Chern class homomorphism** and $c_1(L)$ is the **first Chern class** or **integral Chern class** of the line bundle $L \in \mathrm{Pic}(X)$.

We remark that (see GRIFFITHS and HARRIS [39, p 139])

$$c_1(L_1 \otimes L_2) = c_1(L_1) + c_1(L_2) \text{ and } c_1(L^*) = -c_1(L).$$

If $f : X \to Y$ is a holomorphic map of complex manifolds, then for the pullback

$$c_1(f^*L) = f^*c_1(L).$$

For $X = \mathbf{C}^n$ we get $H^1(X, \mathcal{O}) = 0$ because of Theorem B for Stein manifolds and $H^2(X, \mathbf{Z}) = 0$ so that $H^1(X, \mathcal{O}^*) = 0$. Therefore every line bundle on $X = \mathbf{C}^n$ is analytically trivial.

2.1.1 Proposition

Let $\Lambda \subset \mathbf{C}^n$ be a discrete subgroup and $X = \mathbf{C}^n/\Lambda$. Then there exists a canonical isomorphism

$$H^1(\Lambda, \mathcal{H}^*(\mathbf{C}^n)) \to \mathrm{Pic}(X),$$

which maps a class of cobordant automorphic factors to a holomorphic isomorphism class of line bundles on X.

Proof

Let $\alpha := \{\alpha_\lambda : \lambda \in \Lambda\}$ be an automorphic factor for Λ. Now Λ acts on the trivial bundle $\mathbf{C}^n \times \mathbf{C}$ by

$$\lambda(z,w) := (z+\lambda, \alpha_\lambda(z)w) \qquad (z \in \mathbf{C}^n,\ w \in \mathbf{C}).$$

Then $L_\alpha := (\mathbf{C}^n \times \mathbf{C})/\Lambda$ defines a fibre bundle over $\mathbf{C}^n/\Lambda$ with trivial pullback $\pi^* L_\alpha \simeq \mathbf{C}^n \times \mathbf{C}$ given by the same automorphic factor.
The map

$$\alpha_z : Z^1(\Lambda, \mathcal{H}^*(\mathbf{C}^n)) \to \mathrm{Pic}(X)$$

defined by $\alpha_z(\alpha) := L_\alpha$ is a homomorphism which induces the injective homomorphism

$$\alpha_H : H^1(\Lambda, \mathcal{H}^*(\mathbf{C}^n)) \to \mathrm{Pic}(X).$$

Now let $L \in \mathrm{Pic}(X)$. Then the pullback $\pi^* L$ is analytically trivial. Let $\iota : \pi^* L \to \mathbf{C}^n \times \mathbf{C}$ be such a trivialization with $\iota(z,w) = (z, h_\iota(z)w)$, where $h_\iota \in \mathcal{H}^*(\mathbf{C}^n)$. Then

$$\alpha_\lambda(z) := h_\iota(z+\lambda)h_\iota^{-1}(z) \qquad (z \in \mathbf{C}^n,\ z \in \Lambda)$$

defines an automorphic factor with $\alpha_z(\alpha) = L_\alpha$ so that α_H is surjective. *Q.E.D.*

A holomorphic (differentiable, continuous) function $s : \mathbf{C}^n \to \mathbf{C}$ is called a holomorphic (differentiable, continuous) **automorphic form** belonging to an automorphic factor α_λ if

$$s(z+\lambda) = \alpha_\lambda(z)s(z) \qquad (z \in \mathbf{C}^n,\ \lambda \in \Lambda).$$

Let $\mathcal{A}_\alpha$ be the $\mathbf{C}$–vectorspace of all automorphic forms belonging to α.

2.1.2 Proposition

Let $\Lambda \subset \mathbf{C}^n$ be discrete and L_α be a line bundle on $X = \mathbf{C}^n/\Lambda$ defined by an automorphic factor α_λ ($\lambda \in \Lambda$). Then there exists a canonical isomorphism

$$H^0(X, L_\alpha) \to \mathcal{A}_\alpha.$$

Proof
Let $\sigma : X \to L_\alpha$ be a holomorphic (differentiable, continuous) section. Then the pullback $\pi^*\sigma : \mathbf{C}^n \to \pi^* L$ and a trivialization $\iota : \pi^* L \to \mathbf{C}^n \times \mathbf{C}$ with $\iota(z,w) := (z, h_\iota(z)w)$ define the function $s : \mathbf{C}^n \to \mathbf{C}$ by

$$s(z) := h_\iota(z)\sigma(\pi(z)) \qquad (z \in \mathbf{C}^n).$$

Indeed, by $\alpha_\lambda(z) = h_\iota(z+\lambda)(h_\iota(z))^{-1}$ as in Proposition 2.1.1 we get

$$s(z+\lambda) = h_\iota(z+\lambda)\sigma(\pi(z)) = \alpha_\lambda(z)h_\iota(z)\sigma(\pi(z)) = \alpha_\lambda(z)s(z) \qquad (z \in \mathbf{C}^n)$$

for every $\lambda \in \Lambda$.
Obviously the map $h_\alpha(\sigma) := s$ is an injective homomorphism. On the other hand let $s \in \mathcal{A}_\alpha$. Then $\sigma(\pi(z)) := s(z)(h_\iota(z))^{-1}$ is well defined, and σ is a section of L_α. Then $h_\sigma : H^0(X, L_\alpha) \to \mathcal{A}_\alpha$ is an isomorphism. *Q.E.D.*

The characteristic decomposition of an exponential system

Let $\Lambda \subset \mathbf{C}^n$ be a discrete subgroup.
Every automorphic factor α_λ ($\lambda \in \Lambda$) gives rise to a not uniquely determined **exponential system** of functions

$$a_\lambda \text{ with } \alpha_\lambda := \mathbf{e}(a_\lambda) \quad (\lambda \in \Lambda)$$

where $\mathbf{e}(z) = \exp(2\pi i z)$ ($z \in \mathbf{C}$).
Automatically the cocycle condition for α_λ induces

$$(*) \qquad a_{\lambda+\mu}(z) \equiv a_\mu(z+\lambda) + a_\lambda(z) \mod \mathbf{Z} \qquad (z \in \mathbf{C}^n).$$

The exponential system is an **automorphic summand**, if the **additive cocycle condition**

$$a_{\lambda+\mu}(z) = a_\mu(z+\lambda) + a_\lambda(z) \qquad (z \in \mathbf{C}^n)$$

holds.
Two exponential systems $a_\lambda, \tilde{a}_\lambda$ ($\lambda \in \Lambda$) are **cobordant**, if their associated automorphic factors are cobordant, iff there exists a function $h \in \mathcal{H}(\mathbf{C}^n)$ with

$$\tilde{a}_\lambda(z) \equiv h(z+\lambda) + a_\lambda(z) - h(z) \mod \mathbf{Z} \qquad (\lambda \in \Lambda).$$

They are **strictly cobordant exponential systems**, if

$$\tilde{a}_\lambda(z) = h(z+\lambda) + a_\lambda(z) - h(z) \qquad (z \in \mathbf{C}^n).$$

Let $\Lambda = \mathbf{Z}^n \oplus \Gamma$ be a discrete subgroup and let $X = \mathbf{C}^n/\Lambda$. Then the exponential map $\mathbf{e} : \mathbf{C}^n \to \mathbf{C}^{*n}$ induces an isomorphism

$$H^1(\Lambda, \mathcal{H}^*(\mathbf{C}^n)) \to H^1(\Gamma, \mathcal{H}^*(\mathbf{C}^{*n}))$$

so that

$$H^1(\Gamma, \mathcal{H}^*(\mathbf{C}^{*n})) \to \operatorname{Pic}(X)$$

becomes an isomorphism in the same way as in Proposition 2.1.1. Instead of working with the whole automorphic factor α_λ ($\lambda \in \Lambda$) we can reduce our considerations to the $\mathbf{Z}^n$–periodic $\alpha_\gamma \in \mathcal{H}^*(\mathbf{C}^{*n})$ ($\gamma \in \Gamma$) with cocycle condition for $\gamma, \gamma' \in \Gamma$ only.

Let $\Lambda := \mathbf{Z}^n \oplus \Gamma$ be a discrete subgroup of $\mathbf{C}^n$.
A Γ**–reduced automorphic factor** is a set of holomorphic non–vanishing $\mathbf{Z}^n$–periodic functions $\alpha_\gamma \in \mathcal{H}^*(\mathbf{C}^n)$ ($\gamma \in \Gamma$) with cocycle property for all $\gamma, \gamma' \in \Gamma$.

A **Γ–reduced automorphic summand** is a set of holomorphic $\mathbf{Z}^n$–periodic functions $a_\gamma \in \mathcal{H}(\mathbf{C}^n)$ $(\gamma \in \Gamma)$ with additive cocycle property for all $\gamma, \gamma' \in \Gamma$.
The following decomposition theorem was introduced by KOPFERMANN in 1964 in the case of missing wild summands [64].

2.1.3 The characteristic decomposition of an exponential system (KOPFERMANN)
Let $\Lambda = \mathbf{Z}^n \oplus \Gamma \subset \mathbf{C}^n$ be a lattice. Then every exponential system a_γ $(\gamma \in \Gamma)$ belonging to a Γ–reduced automorphic factor has the **characteristic decomposition**

$$a_\gamma(z) = \langle \chi_\gamma, z \rangle + \frac{1}{2}\langle \chi_\gamma, \gamma \rangle + s_\gamma(z) + c_\gamma + i\, d_\gamma \qquad (\gamma \in \Gamma)$$

with the following properties:

1. $\chi : \Gamma \to \mathbf{Z}^n$ is a homomorphism which defines the alternating bilinear form $A : \Gamma \times \Gamma \to \mathbf{Z}$ by $A(\gamma, \gamma') := \langle \chi_\gamma, \gamma' \rangle - \langle \chi_{\gamma'}, \chi \rangle$ where $\chi_\gamma := \chi(\gamma)$ $(\gamma \in \Gamma)$.
We say that χ is the **Γ–reduced characteristic homomorphism,**
$\langle \chi_\gamma, z \rangle$ $(z \in \mathbf{C}^n)$ the **Γ–reduced characteristic linear form** and
A the **Γ–reduced characteristic alternating bilinear form.**
2. $s_\gamma \in \mathcal{H}(\mathbf{C}^n)$ $(\gamma \in \Gamma)$ is a Γ–reduced automorphic summand with vanishing 0–coefficients of its Fourier expansion,
s_γ is called the **Γ–reduced wild summand** of a_γ.
3. $c_\gamma \in \mathbf{R}$ $(\gamma \in \Gamma)$ are constants with

$$c_{\gamma+\gamma'} \equiv c_\gamma + c_{\gamma'} + \frac{1}{2}A(\gamma, \gamma') \mod \mathbf{Z} \qquad (\gamma, \gamma' \in \Gamma)$$

and $d : \Gamma \to \mathbf{R}$ with $d_\gamma := d(\gamma)$ is a homomorphism.

Uniqueness. The decomposition of a_γ into a linear form $\langle \chi_\gamma, z \rangle$, a Γ–reduced automorphic summand s_γ with a vanishing 0–coefficient and a constant is unique such that c_γ, d_γ are unique.
Cobordism. Cobordant exponential systems have the same characteristic and $d : \Gamma \to \mathbf{R}$ homomorphism. The wild summands are strictly cobordant with a $\mathbf{Z}^n$–periodic function, and the real c_γ change as

$$\tilde{c}_\gamma \equiv c_\gamma + \langle m, \gamma \rangle \mod \mathbf{Z} \qquad (\gamma \in \Gamma) \quad \text{with } m \in \mathbf{Z}^n.$$

Remark. The characteristic homomorphism χ and the characteristic bilinear form A depend only on the line bundle L defined by the automorphic factor $\alpha_\gamma := \mathbf{e}(a_\gamma)$ $(\gamma \in \Gamma)$.
Proof
Let a_γ $(\gamma \in \Gamma)$ be an exponential system for α_γ $(\gamma \in \Gamma)$, which is not necessarily $\mathbf{Z}^n$–periodic. Because the α_γ are $\mathbf{Z}^n$–periodic we get

$$\chi_{\gamma,j} := a_\gamma(z+e_j) - a_\gamma(z) \in \mathbf{Z} \qquad (z \in \mathbf{C}^n)$$

with unit vectors e_j $(j = 1, \cdots, n)$ so that

$$^t\chi_\gamma := (\chi_{\gamma,1}, \cdots, \chi_{\gamma,n}) \in \mathbf{Z}^n.$$

Obviously $f_\gamma(z) := a_\gamma(z) - \langle \chi_\gamma, z \rangle$ is $\mathbf{Z}^n$–periodic such that

$$a_\gamma(z) = \langle \chi_\gamma, z \rangle + f_\gamma(z)$$

is the sum of a linear form and a $\mathbf{Z}^n$–periodic function. Such a decomposition is unique.

Now let $f_\gamma(z) = \sum_{\sigma \in \mathbf{Z}^n} f_\gamma^{(\sigma)} \mathbf{e}(\langle \sigma, z \rangle)$ be the Fourier expansion of f_γ. Set $k_\gamma := f_\gamma^{(0)}$ and $s_\gamma := f_\gamma - k_\gamma$ $(\gamma \in \Gamma)$ such that

$$f_\gamma(z) = s_\gamma(z) + k_\gamma.$$

The cocycle condition for a_γ implies

$$a_{\gamma+\gamma'}(z) \equiv a_{\gamma'}(z+\gamma) + a_\gamma(z) \mod \mathbf{Z}$$

and therefore

$$\begin{aligned}\langle \chi_{\gamma+\gamma'}, z \rangle + s_{\gamma+\gamma'}(z) + k_{\gamma+\gamma'} \equiv \\ \langle \chi_\gamma, z \rangle + \langle \chi_{\gamma'}, z \rangle + s_\gamma(z) + s_{\gamma'}(z+\gamma) + \langle \chi_{\gamma'}, \gamma \rangle + k_\gamma + k_{\gamma'} \quad \mod \mathbf{Z}.\end{aligned}$$

Now as before the linear part is unique such that

$$\langle \chi_{\gamma+\gamma'}, z \rangle = \langle \chi_\gamma, z \rangle + \langle \chi_{\gamma'}, z \rangle.$$

Thus, $\chi : \Gamma \to \mathbf{Z}^n$ becomes a homomorphism. The part of the Fourier expansion without 0–coefficients is unique so that the cocycle condition

$$s_{\gamma+\gamma'}(z) = s_\gamma(z) + s_{\gamma'}(z+\gamma) \qquad (z \in \mathbf{C}^n)$$

is fulfilled. The set s_γ $(\gamma \in \Gamma)$ is an automorphic summand. Finally we get

$$k_{\gamma+\gamma'} \equiv k_\gamma + k_{\gamma'} + \langle \chi_{\gamma'}, \gamma \rangle \mod \mathbf{Z}.$$

Let $c_\gamma + i\, d_\gamma := k_\gamma - \frac{1}{2}\langle \chi_\gamma, \gamma \rangle$ with real c_γ, d_γ. Then by the previous congruence we get

$$c_{\gamma+\gamma'} \equiv c_\gamma + c_{\gamma'} + \frac{1}{2} A(\gamma, \gamma') \mod \mathbf{Z}.$$

By changing γ to γ' we see $A(\gamma, \gamma') \in \mathbf{Z}$.
Finally $d_{\gamma+\gamma'} = d_\gamma + d_{\gamma'}$.
Cobordism. Let $a_\gamma, \tilde{a}_\gamma$ be cobordant exponential systems. We have

$$\tilde{a}_\gamma(z) \equiv h(z+\gamma) + a_\gamma(z) - h(z) \mod \mathbf{Z}$$

with $h \in \mathcal{H}(\mathbf{C}^n)$ such that $\mathbf{e}(h)$ is $\mathbf{Z}^n$–periodic. As above we get a decomposition

$$h(z) = \langle m, z\rangle + f(z)$$

with $m \in \mathbf{Z}^n$ and a $\mathbf{Z}^n$–periodic function f such that

$$\tilde{a}_\gamma(z) \equiv a_\gamma(z) + \langle m, \gamma\rangle + f(z+\gamma) - f(z) \mod \mathbf{Z}.$$

Because of uniqueness the characteristic homomorphisms are the same. The wild summands $s_\gamma, \tilde{s}_\gamma$ of $a_\gamma, \tilde{a}_\gamma$, respectively, transform as

$$\tilde{s}_\gamma(z) = f(z+\gamma) + s_\gamma(z) - f(z).$$

This equation is strict because we use only Fourier-series with vanishing 0–coefficients. The rest is clear. *Q.E.D.*

2.1.4 Extension of Γ–reduced exponential systems

The $\mathbf{Z}^n$–periodic automorphic factor α_λ $(\lambda \in \Lambda)$ with $\Lambda = \mathbf{Z}^n \bigoplus \Gamma$ is defined by $\alpha_{m+\gamma} := \alpha_\gamma$ $(m \in \mathbf{Z}^n,\ \gamma \in \Gamma)$, similar its exponential system $a_{m+\gamma} := a_\gamma$ $(m \in \mathbf{Z}^n,\ \gamma \in \Gamma)$. We can restrict our considerations to exponential systems with $a_m = 0\,(m \in \mathbf{Z}^n)$.

By uniqueness of the characteristic decomposition we see that $\chi_{m+\gamma} = \chi_\gamma$ remains a homomorphism $\chi : \Lambda \to \mathbf{Z}^n$ and $s_{m+\gamma} = s_\gamma$ remains an automorphic summand s_λ $(\lambda \in \Lambda)$ and $d_{m+\gamma} := d_\gamma$ a homomorphism $d : \Lambda \to \mathbf{R}$, but

$$c_{m+\gamma} = c_\gamma - \frac{1}{2}\langle \chi_\gamma, m\rangle \qquad (m \in \mathbf{Z}^n,\ \gamma \in \Gamma).$$

We extend the **characteristic alternating bilinear form** A from $\Gamma \times \Gamma$ to $\Lambda \times \Lambda$ in the same way as it is defined in the decomposition theorem by

$$A(\lambda, \lambda') = \langle \chi_\lambda, \lambda'\rangle - \langle \chi_{\lambda'}, \lambda\rangle \qquad (\lambda, \lambda' \in \Lambda).$$

Because the characteristic homomorphism χ depends only on the line bundle L on $X = \mathbf{C}^n/\Lambda$ the alternating bilinear form A does the same.
For the extended c_λ $(\lambda \in \Lambda)$ we get the same relation

$$c_{\lambda+\lambda'} \equiv c_\lambda + c_{\lambda'} + \frac{1}{2}A(\lambda, \lambda') \mod \mathbf{Z} \qquad (\lambda, \lambda' \in \Lambda).$$

The **integral defect** of the cocycle relation for the extended exponential system a_λ $(\lambda \in \Lambda)$ is

$$D(\lambda, \lambda') := a_{\lambda+\lambda'}(z) - a_{\lambda'}(z+\lambda) - a_\lambda(z) \in \mathbf{Z} \qquad (\lambda, \lambda' \in \Lambda).$$

So we can easily prove by direct calculation that the extended

$$A(\lambda,\lambda') = D(\lambda,\lambda') - D(\lambda',\lambda) = a_\lambda(z+\lambda') + a_{\lambda'}(z) - a_{\lambda'}(z+\lambda) - a_\lambda(z)$$

remains integral on $\Lambda \times \Lambda$.

2.1.5 Lemma
Let $\Lambda \subset \mathbf{C}^n$ be a discrete subgroup of $\mathbf{C}^n$ and $X = \mathbf{C}^n/\Lambda$. Then there exists a canonical isomorphism

$$H^2(X,\mathbf{Z}) \to \mathrm{Alt}^2(\Lambda,\mathbf{Z}),$$

which maps especially the integral Chern class $c_1(L)$ of a line bundle L on X defined by an automorphic factor with exponential system a_λ ($\lambda \in \Lambda$) to the characteristic alternating form

$$A(\lambda,\lambda') = a_\lambda(z+\lambda') + a_{\lambda'}(z) - a_{\lambda'}(z+\lambda) - a_\lambda(z) \qquad (\lambda,\lambda' \in \Lambda)$$

For the *proof* see LANGE–BIRKENHAKE [66] pp 25/26. It works for any discrete subgroup $\Lambda \in \mathbf{C}^n$, not only with rank $2n$.

Automorphic summands

In contrast to the case of the torus wild summands need not be cobordant to 0 in general.

Let $\Lambda = \mathbf{Z}^n \oplus \Gamma$ be a toroidal lattice and s_γ ($\gamma \in \Gamma$) a Γ–reduced automorphic summand with Fourier series expansion

$$s_\gamma(z) = \sum_{\sigma\in\mathbf{Z}^n\setminus\{0\}} s_\gamma^{(\sigma)}\mathbf{e}(\langle\sigma,z\rangle) \qquad (z \in \mathbf{C}^n).$$

The irrationality condition (I) for standard coordinates garanties that for every $\sigma \in \mathbf{Z}^n\setminus\{0\}$ there exists a γ_{j_σ} of a fixed basis $\gamma_1,\cdots,\gamma_q$ of Γ such that $\langle\sigma,\gamma_{j_\sigma}\rangle \notin \mathbf{Z}$. The cocycle condition implies

$$s_\gamma(z+\gamma_{j_\sigma}) - s_\gamma(z) = s_{\gamma_{j_\sigma}}(z+\gamma) - s_{\gamma_{j_\sigma}}(z) \qquad (z \in \mathbf{C}^n),$$

such that for the Fourier coefficients of an automorphic summand

$$s_\gamma^{(\sigma)}[\mathbf{e}(\langle\sigma,\gamma_{j_\sigma}\rangle) - 1] = s_{\gamma_{j_\sigma}}^{(\sigma)}[\mathbf{e}(\langle\sigma,\gamma\rangle) - 1]$$

holds with $\mathbf{e}(\langle\sigma,\gamma_{j_\sigma}\rangle) - 1 \neq 0$.
We want to construct a Fourier series

$$h(z) = \sum_{\sigma\in\mathbf{Z}^n\setminus\{0\}} h^{(\sigma)}\mathbf{e}(\langle\sigma,z\rangle)$$

with

$$(*)\qquad h(z+\gamma)-h(z)=s_\gamma(z)$$

on $\mathbf{MC}_\Lambda$ or even $\mathbf{C}^n$ so that s_γ becomes cobordant to 0 on $\mathbf{MC}_\Lambda$ or $\mathbf{C}^n$, respectively. If it is true on the $\mathbf{C}^n$, then the $h^{(\sigma)}$ are uniquely defined by

$$h^{(\sigma)}[\mathbf{e}(\langle\sigma,\gamma_{j_\sigma}\rangle)-1]=s^{(\sigma)}_{\gamma_{j_\sigma}}\qquad(\sigma\in\mathbf{Z}^n\setminus\{0\}).$$

But in contrary to the classical torus case h does not converge in general. We shall see that the wild summands in general are cobordant to 0 only on the maximal complex subspace $\mathbf{MC}_\Lambda$ of $\mathbf{R}_\Lambda$, as VOGT proved [115, 116].

We remark first that a Fourier series $\sum_{\sigma\in\mathbf{Z}^n} h^{(\sigma)}\mathbf{e}(\langle\sigma,z\rangle)$ converges everywhere in $\mathbf{C}^n$, iff the associated real Laurent series $\sum_{\sigma\in\mathbf{Z}^n}\left|h^{(\sigma)}\right|k^\sigma$ converges for every $k\in\mathbf{R}^n_{>0}$.

2.1.6 Proposition (VOGT)
Let $X=\mathbf{C}^n/\Lambda$ be a toroidal group of type q. Then

a) *Every* automorphic summand is cobordant to an automorphic summand constant on the maximal $\mathbf{C}$–linear subspace $\mathbf{MC}_\Lambda$ of $\mathbf{R}_\Lambda$.
b) Let $G:=(\gamma_1,\cdots,\gamma_q)$ be the basis of Γ where $\Lambda=\mathbf{Z}^n\oplus\Gamma$.
Then *every* automorphic summand is cobordant to a constant, iff there exists a real $r>0$ such that

$$\textbf{(TS)}\qquad r^{-|\sigma|}\le \operatorname{dist}({}^tG\sigma,\mathbf{Z}^q)=\inf_{\tau\in\mathbf{Z}^q}\left|{}^tG\sigma-\tau\right|\qquad(\sigma\in\mathbf{Z}^n\setminus\{0\}).$$

c) Let R be the glueing matrix of Λ in toroidal coordinates.
Then *every* automorphic summand is cobordant to a constant, iff there exists a real $r>0$ such that

$$\textbf{(TT)}\qquad r^{-|\sigma|}\le \operatorname{dist}({}^tR\sigma,\mathbf{Z}^{2q})=\inf_{\tau\in\mathbf{Z}^{2q}}\left|{}^tR\sigma-\tau\right|\qquad(\sigma\in\mathbf{Z}^{n-q}\setminus\{0\}).$$

Proof
As shown in 1.1.12 we can change the standard coordinates $w=\binom{u}{v}$ where Λ has the basis (I_n,G) with a refined linear transformation

$$l(u,v)=((\mathrm{Im}\hat{T})^{-1}u,v+R_1u)\qquad(u\in\mathbf{C}^q,\ v\in\mathbf{C}^{n-q})$$

into toroidal coordinates z where Λ has the basis

$$P=\begin{pmatrix}0 & B_1 & B_2\\ I_{n-q} & R_1 & R_2\end{pmatrix}.$$

$\hat{T}$ is the square matrix of the first q rows of G with invertible Im$\hat{T}$, $B := (B_1, B_2)$ with Im $B_2 = I_q$ the basis of a torus and $R := (R_1, R_2)$ the real glueing matrix. The maximal $\mathbf{C}$–linear subspace $\mathbf{MC}_\Lambda$ of $\mathbf{R}_\Lambda$ becomes the space of the first q variables.

In coordinates z after refined transformations from standard coordinates we want to get an $l(\mathbf{Z}^n)$–periodic automorphic summand $\tilde{s}_\gamma$ $(\gamma \in \Gamma)$ cobordant to s_γ $(\gamma \in \Gamma)$. We want to find a holomorphic function h such that

$$\tilde{s}_\gamma(z) = h(z + l(\gamma)) - s_\gamma(z) + h(z) \qquad (\gamma \in \Gamma)$$

is constant on $\mathbf{MC}_\Lambda$ or on $\mathbf{C}^n$, respectively. This means

$$\frac{\partial h}{\partial z_j}(z + l(\gamma)) - \frac{\partial h}{\partial z_j}(z) = -\frac{\partial s_\gamma}{\partial z_j}(z) \qquad (\gamma \in \Gamma)$$

for $j = 1, \cdots, p$, where $p = q$ or $= n$, respectively. For this purpose it is sufficient to prove the

Statement. There are $l(\mathbf{Z}^n)$–periodic entire functions h_j with

$$(**) \qquad h_j(z + l(\gamma)) - h_j(z) = -\frac{\partial s_\gamma}{\partial z_j}(z) \qquad (\gamma \in \Gamma)$$

for $j = 1, \cdots, p$, namely in case a) for $p = q$ in general and in the cases b) and c) for $p = n$, if the given irrationality conditions (TS) or (TT), respectively, hold.

Indeed. If the statement is true, the function

$$h(z) := \sum_{j=1}^{p} \int_0^1 h_j(tz_1, \cdots, tz_p, z_{p+1}, \cdots, z_n)dt \ z_j$$

is $l(\mathbf{Z}^n)$–periodic, $\frac{\partial h}{\partial z_j} = h_j$ $(j = 1, \cdots, p)$ so that

$$\tilde{s}_\gamma(z) := h(z + l(\gamma)) + s_\gamma(z) - h(z) \qquad (\gamma \in \Gamma)$$

defines an $l(\mathbf{Z}^n)$–periodic automorphic summand constant in the first q variables.

Proof of the statement. After linear change l^{-1} from refined toroidal coordinates z to standard coordinates w we get the Fourier series

$$s_\gamma \circ l(w) = \sum_{\sigma \in \mathbf{Z}^n \setminus \{0\}} s_\gamma^{(\sigma)} \mathbf{e}(\langle \sigma, w \rangle)$$

and

$$h \circ l(w) = \sum_{\sigma \in \mathbf{Z}^n \setminus \{0\}} h_j^{(\sigma)} \mathbf{e}(\langle \sigma, w \rangle) \qquad (\gamma \in \Gamma),$$

such that $(**)$ is equivalent to

$$h_j^{(\sigma)}[\mathbf{e}(\langle\sigma,\gamma\rangle)-1]=-2\pi i s_\gamma^{(\sigma)}\langle\sigma,l^{-1}(e_j)\rangle \qquad (\gamma\in\Gamma,\ \sigma\in\mathbf{Z}^n\setminus\{0\})$$

for $j=1,\cdots,p$ where $p=q$ or $p=n$, respectively.
Now for every $\sigma\in\mathbf{Z}^n\setminus\{0\}$ we choose a $\gamma=\gamma_{j_\sigma}$ of a fixed basis $(\gamma_1,\cdots,\gamma_q)$ of Γ such that

$$(***) \qquad |\mathbf{e}(\langle\sigma,\gamma_{j_\sigma}\rangle)-1|\geq|\mathbf{e}(\langle\sigma,\gamma_j\rangle)-1| \qquad (j=1,\cdots,p).$$

Besides this we remember 1.13 that after changing back from refined toroidal coordinates to standard coordinates

$$(\dagger) \qquad l^{-1}(e_j)=\operatorname{Im}\gamma_j\ (j=1,\cdots,q) \text{ and } l^{-1}(e_j)=e_j\ (j=q+1,\cdots,n).$$

Case a) We have to show that the Fourier series with coefficients

$$h_j^{(\sigma)}:=-2\pi i s_{\gamma_{j_\sigma}}^{(\sigma)}\langle\sigma,\operatorname{Im}\gamma_j\rangle[\mathbf{e}(\langle\sigma,\gamma_{j_\sigma}\rangle)-1]^{-1}$$

converges everywhere $(j=1,\cdots,q)$. For that we first fix a real $\varepsilon>0$ and define

$$J_{\geq\varepsilon}:=\{\sigma\in\mathbf{Z}^n:|\mathbf{e}(\langle\sigma,\gamma_{j_\sigma}\rangle)-1|\geq\varepsilon\}.$$

Then

$$\sum_{\sigma\in J_{\geq\varepsilon}}|s_{\gamma_{j_\sigma}}||\langle\sigma,\operatorname{Im}\gamma_j\rangle||\mathbf{e}(\langle\sigma,\gamma_{j_\sigma}\rangle)-1|^{-1}k^\sigma\leq\frac{1}{\varepsilon}\sum_{\sigma\in J_{\geq\varepsilon}}|s_{\gamma_{j_\sigma}}||\langle\sigma,\operatorname{Im}\gamma_j\rangle|\,k^\sigma$$

is obviously convergent for every $k\in\mathbf{R}^n_{\geq 0}$ because the series $\partial s_\gamma/\partial z_j$ $(j=1,\cdots,q)$ converge everywhere.
Now define

$$J_{<\varepsilon}^{(j)}:=\{\sigma\in\mathbf{Z}^n:\ 0<|\mathbf{e}(\langle\sigma,\gamma_j\rangle)-1|<\varepsilon\} \qquad (j=1,\cdots,q).$$

The map $x\mapsto\exp x$ $(x\in\mathbf{R})$ is a diffeomorphism so that locally

$$|\langle\sigma,u\rangle|\leq c\,|\exp(\langle\sigma,u\rangle)-1|$$

with a constant $c>0$. Therefore with $(***)$

$$|2\pi\langle\sigma,\operatorname{Im}\gamma_j\rangle|\,|\mathbf{e}(\langle\sigma,\operatorname{Im}\gamma_{j_\sigma}\rangle)-1|^{-1}\leq c_j \qquad (\sigma\in J_{<\varepsilon}^{(j)}),$$

such that

$$\sum_{\sigma\in J_{<\varepsilon}^{(j)}}\left|s_{\gamma_{j_\sigma}}^{(\sigma)}\right||\langle\sigma,\operatorname{Im}\gamma_j\rangle|\,|\mathbf{e}(\langle\sigma,\gamma_{j_\sigma}\rangle)-1|^{-1}\,k^\sigma$$

$$\leq\sum\frac{c_j}{2\pi}\left|s_{\gamma_{j_\sigma}}^{(\sigma)}\right|k^\sigma \qquad (j=1,\cdots,q).$$

These series converge because the series of s_{γ_j} do $(j=1,\cdots,q)$.

Case b) We consider the Fourier series with coefficients

$$h_k^{(\sigma)} = -2\pi i s_{\gamma_{j_\sigma}}^{(\sigma)} \langle \sigma, e_k \rangle [\mathbf{e}(\langle \sigma, \gamma_{j_\sigma} \rangle) - 1]^{-1} \qquad (k = q+1, \cdots, n).$$

As in a) the series with coefficients $\sigma \in J_{>\varepsilon}$ converge so that we have to study the series with coefficients

$$J_{<\varepsilon} := \{\sigma \in \mathbf{Z}^n : \ 0 < |\mathbf{e}(\langle \sigma, \gamma_{j_\sigma} \rangle) - 1| < \varepsilon\}.$$

The map $z \mapsto \mathbf{e}(z)$ as map $\mathbf{C}/\mathbf{Z} \to \mathbf{C}^*$ and its inverse are diffeomorphisms so that locally

$$d_j \, |\mathbf{e}(\langle \sigma, \gamma_{j_\sigma} \rangle) - 1| \le \operatorname{dist}(\langle \sigma, \gamma_j \rangle, \mathbf{Z}) \le c_j \, |\mathbf{e}(\langle \sigma, \gamma_j \rangle) - 1| \qquad (\sigma \in J_{<\varepsilon})$$

with suitable $d_j, c_j > 0$ and

$$\operatorname{dist}(\langle \sigma, \gamma_j \rangle, \mathbf{Z}) = \inf_{\tau_j \in \mathbf{Z}} |\langle \sigma, \gamma_j \rangle - \tau_j| \qquad (j = 1, \cdots, q).$$

If the irrationality condition (TS) is fulfilled, there exists a natural $N > 0$ with

$$N^{-|\sigma|} \le \inf_{\tau \in \mathbf{Z}^q} |{}^t G \sigma - \tau| = \operatorname{dist}({}^t G \sigma, \mathbf{Z}^q) \qquad (\sigma \in \mathbf{Z}^q).$$

Remember that $G = (\gamma_1, \cdots, \gamma_q)$ is the fixed basis of Γ. Then

$$N^{-|\sigma|} \le \sqrt{n} \inf_{\tau_{j_\sigma}} |\langle \sigma, \gamma_{j_\sigma} \rangle - \tau_{j_\sigma}| \le c \, |\mathbf{e}(\langle \sigma, \gamma_{j_\sigma} \rangle) - 1|$$

with a suitable $c > 0$ such that

$$\sum_{\sigma \in J_{<\varepsilon}} \left| s_{\gamma_{j_\sigma}}^{(\sigma)} \right| |\langle \sigma, e_k \rangle| \, |\mathbf{e}(\langle \sigma, \gamma_{j_\sigma} \rangle) - 1|^{-1} k^\sigma \le c \sum \left| s_{\gamma_{j_\sigma}}^{(\sigma)} \right| |\langle \sigma, e_k \rangle| \, N^{|\sigma|} k^\sigma$$

is convergent because the series of $\partial s_{\gamma_j} / \partial z_k$ converge $(k = q+1, \cdots, n)$.
If the irrationality condition (TT) is not fulfilled, then for every natural $N > 0$ there exists a $\sigma_N \in \mathbf{Z}^q$ with

$$\inf_{\tau \in \mathbf{Z}^q} |{}^t G \sigma_N - \tau| = \operatorname{dist}({}^t G \sigma_N, \mathbf{Z}^q) < N^{-|\sigma_N|}.$$

We can choose the σ_N pairwise different. Then there exists a constant $d > 0$ such that

$$d \, |\mathbf{e}(\langle \sigma_N, \gamma_j \rangle - 1| \le \operatorname{dist}(\langle \sigma_N, \gamma_j \rangle, \mathbf{Z}) < N^{-|\sigma_N|} \qquad (\sigma_N \in J_{<\varepsilon}).$$

We want to show that there exists a divergent Fourier series h such that the Γ-reduced automorphic summand

$$a_\gamma(z) := h(z + \gamma) - h(z) \qquad (\gamma \in \Gamma)$$

is convergent. Because h is uniquely determined by a_γ the summand a_γ cannot be cobordant tó a constant.
Indeed, obviously

$$h(z) := \sum_N \mathbf{e}(\langle \sigma_N, z\rangle)$$

is divergent, but

$$a_{\gamma_j}(z) := h(z+\gamma_j) - h(z) = \sum_N [\mathbf{e}(\langle \sigma_N, \gamma_j\rangle) - 1]\mathbf{e}(\langle \sigma_N, z\rangle)$$

converges everywhere because

$$\sum_N |\mathbf{e}(\langle \sigma_N, \gamma_j\rangle) - 1|\, k^{|\sigma_N|} \le d \sum_N \left(\frac{k}{N}\right)^{|\sigma|_N}$$

is convergent for every $k \in \mathbf{R}^n_{>0}$.

Case c) It was proved in a) that every holomorphic summand is cobordant to an automorphic summand which is constant on MC_Λ and $\mathbf{Z}^{n-q}$–periodic in the last $n-q$ variables in toroidal coordinates. So we can apply b) to all automorphic summands with these properties. But then b) holds, if we replace matrix G by glueing matrix R. *Q.E.D.*

From now on for ***toroidal groups*** we can assume that the wild summands in the characteristic decomposition are constant on the maximal **C**–linear subspace MC_Λ of $\mathbf{R}_\Lambda$.

Theta bundles and topologically trivial line bundles

On a torus every exponential system of an automorphic factor defining a line bundle is cobordant to an exponential system of linear polynomials. In the general toroidal case the wild summands can be reduced to constants only on the maximal complex subspace MC_Λ of $\mathbf{R}_\Lambda$.

2.1.7 Definition

Let $X = \mathbf{C}^n/\Lambda$ with $\Lambda = \mathbf{Z}^n \oplus \Gamma$.
An automorphic factor is a **theta factor** or **linearizable**, if it is given by an exponential system of *linear* polynomials. A line bundle L on X is a **theta bundle** or **linearizable**, if it can be given by a theta factor.

The names *theta factor* and *theta bundle* are used in classical torus theory.

2.1.8 Lemma

For a line bundle L on $X = \mathbf{C}^n/\Lambda$ with $\Lambda = \mathbf{Z}^n \oplus \Gamma$ the follwing conditions are equivalent:

1. L is a theta bundle.
2. The wild summand of the automorphic factor of L is cobordant to constants.

Proof
Consider the characteristic decomposition 2.1.3. *Q.E.D.*

On the other hand we get:

2.1.9 Lemma
For a line bundle L on $X = \mathbf{C}^n/\Lambda$ with $\Lambda = \mathbf{Z}^n \oplus \Gamma$ the following statements are equivalent:

1. L is topologically trivial.
2. The integral Chern class $c_1(L) = 0$.
3. The characteristic bilinear and alternating form $A = 0$.
4. The Γ–reduced characteristic homomorphism $\chi = 0$.
5. The automorphic factor defining L can be given by an automorphic summand.

Proof
2⤖3. By Lemma 2.1.5.
1⤚3. If the line bundle L is topologically trivial, then it has a continuous section without zeros so that by Proposition 2.1.2 a continuous automorphic form $f : \mathbf{C}^n \to \mathbf{C}^*$ exists with

$$f(z+\lambda) = \alpha_\lambda(z) f(z) \qquad (\lambda \in \Lambda).$$

Let a_λ be an exponential system for α_λ $(\lambda \in \Lambda)$ and $f = \mathbf{e}(h)$. Then

$$h(z+\lambda) = a_\lambda(z) + h(z) + n_\lambda \quad \text{with } n_\lambda \in \mathbf{Z} \qquad (\lambda \in \Lambda).$$

We calculate $h(z+\lambda+\lambda') - h(\lambda)$
$= h(z+\lambda+\lambda') - h(z+\lambda) + h(z+\lambda) - h(z)$
$= a_{\lambda'}(z+\lambda) + n_{\lambda'} + a_\lambda(z) + n_\lambda$
$= h(z+\lambda+\lambda') - h(z+\lambda') + h(z+\lambda') - h(z)$
$= a_\lambda(z+\lambda') + n_\lambda + a_{\lambda'}(z) + n_{\lambda'}$.

Then

$$a_\lambda(z+\lambda') + a_{\lambda'}(z) = a_{\lambda'}(z+\lambda) + a_\lambda(z).$$

The characteristic bilinear and alternating $A = 0$.
3⤚4. The restricted $A|_{\Gamma\times\Gamma} = 0$, iff the *commutation property* for the *restricted* characteristic homomorphism $\chi|_\Gamma$

$$\langle \chi_\gamma, \gamma' \rangle = \langle \chi_{\gamma'}, \gamma \rangle \qquad (\gamma, \gamma' \in \Gamma)$$

holds. Then we get for the extended bilinear form

$$A(\lambda, \lambda') = \langle \chi_\gamma, m' \rangle - \langle \chi_{\gamma'}, m \rangle \text{ for all } \lambda = m + \gamma,\ \lambda' = m' + \gamma'.$$

If we take $\gamma = \gamma'$ and $m' - m = e_j$ with unit vector e_j $(j = 1, \cdots, n)$, we see $\chi_\gamma = 0$ $(\gamma \in \Gamma)$.

4≻5. If $\chi = 0$, then L is given by the exponential system

$$a_\gamma(z) = s_\gamma(z) + c_\gamma + i\, d_\gamma \ (\gamma \in \Gamma) \text{ with } c_{\gamma+\gamma'} \equiv c_\gamma + c_{\gamma'} \mod \mathbf{Z}\ (\gamma, \gamma' \in \Gamma).$$

Let $G = (\gamma_1, \cdots, \gamma_q) \in \Gamma$ be a basis. Then

$$\hat{c}_{m_1\gamma_1 + \cdots + m_q\gamma_q} := m_1 c_{\gamma_1} + \cdots + m_q c_{\gamma_q}$$

defines a homomorphism $\hat{c} : \Gamma \to \mathbf{R}$ with

$$\hat{c}_\gamma \equiv c_\gamma \mod \mathbf{Z} \qquad (\gamma \in \Gamma)$$

so that the c_γ can be substituted by $\hat{c}_\gamma$ without changing the Γ–reduced automorphic factor defining L. Then $s_\gamma + \hat{c}_\gamma + i\, d_\gamma$ is a Γ–reduced automorphic summand which can be extended to an automorphic summand a_λ $(\lambda \in \Lambda)$ defining L.

5≻1. As we have seen at the beginning of the proof it is enough to show that there exists a continuous $h : \mathbf{C}^n \to \mathbf{C}$ with

$$h(z + \lambda) - h(z) = a_\lambda(z) \qquad (\lambda \in \Lambda).$$

For that let $(U_j)_{j \in J}$ be a locally finite covering of $X = \mathbf{C}^n/\Lambda$ such that with projection $\pi : \mathbf{C}^n \to \mathbf{C}^n/\Lambda$ we have

$$\pi^{-1}(U_j) = \bigcup_{\lambda \in \Lambda} (V_j + \lambda),$$

where $V_j + \lambda$ $(\lambda \in \Lambda)$ are pairwise disjoint and $\pi : V_j + \lambda \to U_j$ is biholomorphic. Moreover let $\tau_j := (\pi | V_j)^{-1} : U_j \to V_j$ $(j \in J)$ and $\{p_j :\ j \in J\}$ be a partition of units belonging to $(U_j)_{j \in J}$:

$$p_j \geq 0, \operatorname{supp} p_j \subset U_j \text{ and } \sum_j p_j = 1.$$

Then we define the continuous $h_j : \mathbf{C}^n \to \mathbf{C}$ by

$$h_j(z) := \begin{cases} p_j(u) a_\lambda(\tau_j(u)) & \text{for } z = \tau_j(u) + \lambda \\ 0 & \text{elsewhere.} \end{cases}$$

Now a_λ $(\lambda \in \Lambda)$ is a summand of automorphy and $z - \tau_j(u) \in \Lambda$ a period so that by cocycle condition

$$\begin{aligned} h_j(z+\mu) - h_j(z) &= h_j[\tau_j(u) + (z - \tau_j(u)) + \mu] - h_j[\tau_j(u) + (z - \tau_j(u))] \\ &= p_j(u) a_{z-\tau_j(u)+\mu}(\tau_j(u)) - p_j(u) a_{z-\tau_j(u)}(\tau_j(u)) \\ &= p_j(u) a_\mu(\tau_j(u) + z - \tau_j(u)) \\ &= p_j(u) a_\mu(z). \end{aligned}$$

The covering $(U_j)_{j\in J}$ is locally finite. With continuous $h := \sum h_j$ we get finally $h(z+\mu) - h(z) = a_\mu(z)$ $(\mu \in \Lambda)$. *Q.E.D.*

The subgroup

$$\mathrm{Pic}_0(X) \subset \mathrm{Pic}(X),$$

the **Picard group of topologically trivial line bundles**, is isomorphic to the group of classes of cobordant Γ–reduced automorphic summands.

The characteristic decomposition of an automorphic factor implies that every automorphic factor is the product of a theta factor, defined by an exponential system of linear polynomials, and an automorphic factor, defined by an exponential system of a wild automorphic summand. So we get the

2.1.10 Theorem (Vogt [115, 116])
Every line bundle L on a toroidal group $X = \mathbf{C}^n/\Lambda$ is the tensor product of a theta bundle L_ϑ and a topologically trivial line bundle L_0

$$L = L_\vartheta \otimes L_0.$$

In general this decomposition is not unique. To see this we define:
A line bundle L on $X = \mathbf{C}^n/\Lambda$ is **homogeneous** on X, if for every translation $T_a(x) := x + a$ $(x \in X)$ the translated bundle $T_a^* L$ is holomorphically isomorphic to L.

2.1.11 Remark
Let $\Lambda = \mathbf{Z}^n \oplus \Gamma$ and L a line bundle on $X = \mathbf{C}^n/\Lambda$. For a Γ–reduced automorphic factor α_γ $(\gamma \in \Gamma)$ of L with exponential system a_γ $(\gamma \in \Gamma)$ the following conditions are equivalent:

1. The a_γ $(\gamma \in \Gamma)$ are constant.
2. The a_γ $(\gamma \in \Gamma)$ can be substituted by a homomorphism $\hat{a} : \Gamma \to \mathbf{C}$ without changing α_γ $(\gamma \in \Gamma)$.
3. The automorphic factor α_γ $(\gamma \in \Gamma)$ is a homomorphism $\alpha : \Gamma \to \mathbf{C}^*$.

We say that the line bundle L can be defined by the **representation** $a : \Gamma \to \mathbf{C}$ or $\alpha : \Gamma \to \mathbf{C}^*$.
Proof
Indeed, we have to see only $1 \succ 2$. But this can be done in the same way as in the proof of Lemma 2.1.9. Because $a_\gamma = c_\gamma + i\, d_\gamma$ with homomorphism $d : \Gamma \to \mathbf{R}$ and

$$c_{\gamma+\gamma'} \equiv c_\gamma + c_{\gamma'} \mod \mathbf{Z} \qquad (\gamma, \gamma' \in \Gamma)$$

we can take $\hat{c}_\gamma$ as the homomorphism $\mathbf{Z}$–generated by $c_{\gamma_1}, \cdots, c_{\gamma_q}$ on a basis $\gamma_1, \cdots, \gamma_q$ of Γ so that $\hat{a}_\gamma := \hat{c}_\gamma + i\, d_\gamma$ $(\gamma \in \Gamma)$ becomes a homomorphism with the desired property. *Q.E.D.*

VOGT proved in 1982 that a line bundle is homogeneous, iff it is topologically trivial under the assumption that every line bundle on $X = \mathbf{C}^n/\Lambda$ is a theta bundle [116]. ABE proved in 1989 that in any case a line bundle is homogeneous iff it is given by a representation [3].

2.1.12 Theorem
For a line bundle L on a toroidal group the following statements are equivalent:

1. L is homogeneous.
2. L is a topologically trivial theta bundle.
3. L can be defined by a representation.

Proof
Let $\Lambda = \mathbf{Z}^n \oplus \Gamma$.
1≻2. The translated bundle $T_x^* L$ is defined by the exponential system

$$a^*_\gamma(z) := a_\gamma(z+x) = a_\gamma(z) + \langle \chi_\gamma, x\rangle + s_\gamma(z+x) - s_\gamma(z).$$

$T_x^* L$ is holomorphically isomorphic to L, iff the defining Γ–reduced automorphic factors are cobordant with a $\mathbf{Z}^n$–periodic $\mathbf{e}(\tau_x)$ where $\tau_x : \mathbf{C}^n \to \mathbf{C}$ is holomorphic. This means that

$$a_\gamma(z) + \langle \chi_\gamma, x\rangle + s_\gamma(z+x) - s_\gamma(z) \equiv \tau_x(z+\gamma) + a_\gamma(z) - \tau_x(z) \mod \mathbf{Z},$$

and this is equivalent to

$$s_\gamma(z+x) - s_\gamma(z) + \langle \chi_\gamma, x\rangle \equiv \tau_x(z+\gamma) - \tau_x(z) \mod \mathbf{Z} \qquad (\gamma \in \Gamma).$$

Now as in proof of the theorem there exists a unique decomposition

$$\tau_x(z) = \langle m_x, z\rangle + q_x(z)$$

with $m_x \in \mathbf{Z}^n$ and $\mathbf{Z}^n$–periodic q_x such that

$$s_\gamma(z+x) - s_\gamma(z) + \langle \chi_\gamma, x\rangle \equiv \langle m_x, \gamma\rangle + q_x(z+\gamma) - q_x(z) \mod \mathbf{Z}.$$

Such a decomposition is unique so that we get

$$(*) \qquad \langle \chi_\gamma, x\rangle \equiv \langle m_x, \gamma\rangle \mod \mathbf{Z} \qquad (\gamma \in \Gamma)$$

and

$$(**) \qquad s_\gamma(z+x) - s_\gamma(z) = q_x(z+\gamma) - q_x(z).$$

Then $s_\gamma(z+x)$ and s_γ are strictly cobordant.
Let us consider $(*)$

$$\langle \chi_\gamma, x\rangle \equiv \langle m_x, \gamma\rangle \mod \mathbf{Z} \qquad (\gamma \in \Gamma).$$

On toroidal groups $\langle m, \gamma\rangle \equiv \langle n, \gamma\rangle \mod \mathbf{Z}$ $(\gamma \in \Gamma)$ for $m, n \in \mathbf{Z}^n$ implies $m = n$ because of the irrationality condition.
So the map

$$\mathbf{C}^n \ni x \mapsto m_x \in \mathbf{Z}^n$$

by $(*)$ becomes a homomorphism. Suppose that there exists a $\gamma \in \Gamma$ with $\chi_\gamma \neq 0$. Let $a := \chi_\gamma / N$ with N so big that $\langle \chi_\gamma, a\rangle < 1$. Then $\varepsilon a \mapsto m_{\varepsilon a} \neq 0$ for all ε with $0 < \varepsilon < 1$. Our map becomes one to one on the real interval εa $(0 < \varepsilon < 1)$. But this cannot be because $\mathbf{Z}^n$ is countable. Then $\chi = 0$ and L is topologically trivial (Lemma 2.1.9).

Next consider $(**)$

$$s_\gamma(z+x) = q_x(z+\gamma) + s_\gamma(z) - q_x(z)$$

which demonstrates that all the automorphic summands $s_\gamma(z+x)$ are strictly cobordant to s_γ. Remember that a consequence of the cocycle condition for automorphic summands $s_\gamma(z) = \sum s_\gamma^{(\sigma)} \mathbf{e}(\langle \sigma, z\rangle)$ on toroidal groups is

$$s_\gamma^{(\sigma)}[\mathbf{e}(\langle \sigma, \gamma_{j_\sigma}\rangle) - 1] = s_{\gamma_{j_\sigma}}^{(\sigma)}[\mathbf{e}(\langle \sigma, \gamma\rangle) - 1]$$

for every $\sigma \in \mathbf{Z}^n \setminus \{0\}$ with a suitable γ_{j_σ} of a fixed basis $\gamma_1, \cdots, \gamma_q$ of Γ so that $\mathbf{e}(\langle \sigma, \gamma_{j_\sigma}\rangle) - 1 \neq 0$. Then we can define

$$p^{(\sigma)} := s_{\gamma_{j_\sigma}}^{(\sigma)}[\mathbf{e}(\langle \sigma, \gamma_{j_\sigma}\rangle) - 1]^{-1} \qquad (\sigma \in \mathbf{Z}^n \setminus \{0\}).$$

Especially $q(x) := q_x(0)$ has to converge absolutely for all $x \in \mathbf{C}^n$, and by $(**)$

$$q_x(z) = \sum_{\sigma \in \mathbf{Z}^n \setminus \{0\}} p^{(\sigma)}[\mathbf{e}(\langle \sigma, x\rangle) - 1]\, \mathbf{e}(\langle \sigma, z\rangle),$$

in particular

$$q(x) = \sum_{\sigma \in \mathbf{Z}^n \setminus \{0\}} p^{(\sigma)}[\mathbf{e}(\langle \sigma, x\rangle) - 1].$$

Indeed, for any $k \in \mathbf{R}^n_{>1}$ we have

$$k^\sigma \leq \left|2^{|\sigma|} k^\sigma - 1\right| \qquad (\sigma \geq 0, \text{ but } \sigma \neq 0).$$

Then there exists a suitable $x \in \mathbf{C}^n$ with $2^{|\sigma|} k^\sigma = \mathbf{e}(\langle \sigma, x\rangle)$ for all $\sigma \geq 0$, but $\sigma \neq 0$. This shows that the part

$$\sum_{\sigma\geq 0,\sigma\neq 0}\left|p^{(\sigma)}\right|k^{\sigma} \text{ of } \sum_{\sigma\in\mathbf{Z}^n\setminus\{0\}}\left|p^{(\sigma)}\right|k^{\sigma}$$

must be convergent, because the same part of the series for q_x is absolutely convergent. In the same way we can see that the other parts are convergent for some $\sigma_j \leq 0$. So

$$p(x) := \sum_{\sigma\in\mathbf{Z}^n\setminus\{0\}} p^{(\sigma)}\mathbf{e}(\langle\sigma,x\rangle)$$

defines a convergent Fourier series so that $q(x) = p(x) - p(0)$ must be holomorphic everywhere.

Finally condition (**) implies for $z = 0$

$$s_\gamma(x) = q_x(\gamma) + s_\gamma(0) - q_x(0).$$

With $q_x(\gamma) = q(x+\gamma) - q(\gamma)$ and $q_x(0) = q(x)$ we get

$$s_\gamma(x) = q(x+\gamma) + s_\gamma(0) - q(\gamma) - q(x)$$

so that s_γ $(\gamma \in \Gamma)$ is cobordant to zero. The line bundle must be a theta bundle.

2≻3. If the exponential system a_γ defines a topologically trivial theta bundle L, then $\chi = 0$ and $s_\gamma = 0$ so that $a_\gamma := c_\gamma + i\,d_\gamma$ $(\gamma \in \Gamma)$ is constant. By the previous remark L is defined by a representation.

3≻1. If the defining automorphic factor is a representation, the exponential system a_γ must be constant. But then the characteristic decomposition implies $\chi = 0$, $s_\gamma = 0$ $(\gamma \in \Gamma)$ so that a_γ is invariant against translations. *Q.E.D.*

In consequence, the decomposition of a line bundle L on a toroidal group into the tensor product of a theta bundle and a topologically trivial line bundle is unique only up to a homogeneous line bundle.

The Néron–Severi group

$$\mathrm{NS}(X) := \mathrm{Pic}(X)/\mathrm{Pic}_0(X)$$

of a toroidal group $X = \mathbf{C}^n/\Lambda$ with $\Lambda = \mathbf{Z}^n \oplus \Gamma$ can be represented by exponential systems

$$a_\gamma(z) = \langle\chi_\gamma, z\rangle + \frac{1}{2}\langle\chi_\gamma,\gamma\rangle + c_\gamma \qquad (\gamma\in\Gamma),$$

where we have to identify those who differ by a representation. But for any other

$$\hat{a}_\gamma(z) = \langle\chi_\gamma, z\rangle + \frac{1}{2}\langle\chi_\gamma,\gamma\rangle + \hat{c}_\gamma \qquad (\gamma\in\Gamma)$$

the difference defines a representation and vice versa. So for toroidal groups $X = \mathbf{C}^n/\Lambda$ with $\Lambda = \mathbf{Z}^n \oplus \Gamma$

$$\mathrm{NS}(X) \subset \mathrm{Hom}(\Gamma, \mathbf{Z}^n)$$

is a subgroup.

2.2 Cohomology of toroidal groups

The Dolbeault cohomology is comparable with the cohomology of torus groups only for toroidal theta groups. The Dolbeault cohomology groups of toroidal wild groups have non–Hausdorff topology. Toroidal wild groups have differential forms not cohomologous to those with constant coefficients.

Toroidal theta and wild groups

Over a torus every line bundle is linearizable. In the general case we define:

2.2.1 Definition
A toroidal group $X = \mathbf{C}^n/\Lambda$ is a **toroidal theta group**, iff *every* line bundle L on X is a theta bundle otherwise a **toroidal wild group**.

In the section about theta bundles and topologically trivial line bundles we proved that every line bundle L on X is a theta bundle, iff every summand of automorphy is cobordant to constants. By the results of Proposition 2.1.6 and Lemma 2.1.9 we know that this is true, iff some special conditions hold. So we get the

2.2.2 Theorem (Vogt)
For any toroidal group $X = \mathbf{C}^n/\Lambda$ of type q the following statements are equivalent:

1. X is a theta group.
2. Every summand of automorphy of Λ is cobordant to constants.
3. For any basis $G = (\gamma_1, \cdots, \gamma_q)$ of Γ in standard coordinates there exists a positive real r such that

$$\text{(TS)} \qquad r^{-|\sigma|} \leq \operatorname{dist}({}^tG\sigma, \mathbf{Z}^q) = \inf_{\tau \in \mathbf{Z}^q} |{}^tG\sigma - \tau| \qquad (\sigma \in \mathbf{Z}^n \setminus \{0\}),$$

(Condition for theta groups in standard coordinates)

4. For any basis in toroidal coordinates with real glueing matrix R there exists a positive real r such that

$$\text{(TT)} \qquad r^{-|\sigma|} \leq \operatorname{dist}({}^tR\sigma, \mathbf{Z}^{2q}) = \inf_{\tau \in \mathbf{Z}^{2q}} |{}^tR\sigma - \tau| \qquad (\sigma \in \mathbf{Z}^{n-q} \setminus \{0\}).$$

(Condition for theta groups in toroidal coordinates)

With this theorem it is possible to construct toroidal theta groups as well as toroidal wild groups for any dimension $n > 1$ and type q with $0 < q < n$. For that we remember [see 1.1.12] that every toroidal group of type q is a natural principal $\mathbf{C}^{*n-q}$–fibre bundle with a torus group of dimension q as base space.

VOGT constructed in 1981 2–dimensional toroidal theta and toroidal wild groups as $\mathbf{C}^*$–fibre bundles over the onedimensional torus which is generated by the periods $(1, i)$. [115]
KAZAMA and UMENO constructed in 1984 n–dimensional toroidal theta and toroidal wild groups as $\mathbf{C}^*$–fibre bundles over $(n-1)$–dimensional torus groups which are generated by the periods (I_{n-1}, iI_{n-1}). [56]
CAPOCASA and CATANESE proved in 1991 that for every n and q with $0 < q < n$ there exist toroidal wild groups of dimension n and type q. [20]

2.2.3 Proposition

For every n and q with $0 < q < n$ and for every torus group T of complex dimension q there are toroidal theta groups and toroidal wild groups which are natural principal $\mathbf{C}^{*n-q}$–fibre bundles over T.
Proof
Obviously the proposition is equivalent with the construction of glueing matrices

$$R = (\alpha, 0, \cdots, 0) \text{ with } \mathbf{Q}\text{–linearly independent } {}^t(\alpha_1, \cdots, \alpha_{n-q}) = \alpha \in \mathbf{R}^{n-q},$$

such that the property 2.2.2(4) holds for some one and it does not hold for other ones. So we have to give examples with

$$(T) \qquad \exists N \in \mathbf{N}_{>0}\ \forall \sigma \in \mathbf{Z}^{n-q} \setminus \{0\}\ \forall \tau \in \mathbf{Z} : \ N^{-|\sigma|} \leq |\langle \sigma, \alpha \rangle + \tau|$$

and other ones with

$$(W) \qquad \forall N \in \mathbf{N}_{>0}\ \exists \sigma_N \in \mathbf{Z}^{n-q} \setminus \{0\}\ \exists \tau_N \in \mathbf{Z} : |\langle \sigma_N, \alpha \rangle + \tau_N| \leq N^{-|\sigma_N|},$$

in both cases with $\mathbf{Q}$–linearly independent $\alpha_1, \cdots, \alpha_{n-q}$.

ELSNER [29] constructed the following examples:

In the case of *toroidal theta groups* (T) we start with the first $m := n - q$ prime numbers $p_1 := 2, p_2 := 3, \cdots, p_m$ and define

$$\alpha := (\log p_1, \cdots, \log p_m).$$

A theorem of ALAN BAKER [15, Theorem 3.1] guarantees the existence of a real $c_m > 0$ such that

$$(*) \qquad \forall (\sigma, \tau) \in \mathbf{Z}^{m+1} \setminus \{0\} : |\tau \cdot 1 + \sigma_1 \log p_1 + \cdots + \sigma_m \log p_m| > S^{-c_m}$$

with

$$S := 2 \max \{|\tau|, |\sigma_1|, \cdots, |\sigma_m|\} \geq 2.$$

If $|\tau + \langle \sigma, \alpha \rangle| \geq 1$, then there exists nothing to prove.
If $|\tau + \langle \sigma, \alpha \rangle| < 1$, then $|\tau| \leq 1 + |\langle \sigma, \alpha \rangle| \leq 1 + \|\sigma\| \, \|\alpha\|$ with maximum norm.

If $2|\tau| \leq S$, then $S = 2\|\sigma\|$ otherwise $S = 2|\tau|$ so that in any case $S \leq 2(1+m)\|\sigma\|\|\alpha\|$ because $2\|\alpha\| = 2\log p_m \geq 1$ $(m \geq 1)$. With $(*)$ we get

$$|\tau + \langle\sigma, \alpha\rangle| > k_m \|\sigma\|^{-c_m} \geq k_m |\sigma|^{-c_m} \quad \text{where} \quad k_m := [2(1+m)\|\alpha\|]^{-1/c_m}.$$

Finally we can find a sufficiently big $N \in \mathbf{N}$ so that $N^x \geq k_m^{-1} x^{c_m}$ $(x \geq 1)$, especially for $x := |\sigma|$

$$|\tau + \langle\sigma, \alpha\rangle| > N^{-|\sigma|}.$$

Then $\alpha_1 = \log p_1, \cdots, \alpha_m = \log p_m$ must be $\mathbf{Q}$-linearly independent.

In the case of *toroidal wild groups* (W) we start with strictly monotone sequence

$$s_1 := 1, \; s_{\mu+1} := \mu m 2^{m s_\mu} \qquad (\mu \geq 1)$$

with $m := n - q$ and define

$$\alpha_1 := \sum_{\mu=1}^{\infty} \frac{1}{2^{s_\mu}}, \quad \alpha_j := \alpha_1^j \qquad (j = 1, \cdots, m),$$

$$\alpha := (\alpha_1, \cdots, \alpha_m).$$

By

$$\sigma_{N,j} := (2^{s_N})^j \,, \; \tau_{N,j} := \left(\sum_{\mu=1}^{N} 2^{s_N - s_\mu}\right)^j \qquad (j = 1, \cdots, m)$$

$$\sigma_N := (\sigma_{N,1}, \cdots, \sigma_{N,m}), \; \tau_N := \sum_{j=1}^{m} \tau_{N,j},$$

we get

$$(**) \qquad 0 < \alpha_1\sigma_{N,1} - \tau_{N,1} = \sum_{\mu=N+1}^{\infty} 2^{s_N - s_\mu} < 2^{s_N - s_{N+1} + 1} \leq 1 \qquad (N \geq 1).$$

Because $|\alpha_1| < 1, \; 0 < \tau_{N,1} < \sigma_{N,1}$, we have

$$0 < \alpha_j\sigma_{N,j} - \tau_{N,j} = \alpha_1^j\sigma_{N,1}^j - \tau_{N,1}^j \leq j\sigma_{N,1}^{j-1}(\alpha_1\sigma_{N,1} - \tau_{N,1})$$

and

$$\sum_{j=1}^{m} j\sigma_{N,1}^{j-1} \leq m^2\sigma_{N,1}^{m-1}.$$

Then

$$|\langle\alpha, \sigma_N\rangle - \tau_N| \leq m^2 2^{m s_N - s_{N+1} + 1}.$$

Now $x^2 N^x < 2^{Nx}$ $(x \geq 9)$ so that for $x := m2^{m s_N}$ $(N \geq 3)$

$$m^2 2^{m s_N - s_{N+1} + 1} < m^2 2^{2 m s_N - s_{N+1}} < N^{-m 2^{m s_N}}$$

and then

$$|\langle\alpha,\sigma_N\rangle-\tau_N|<N^{-m\sigma_{N,m}}\leq N^{-|\sigma_N|}\qquad(N\geq 3).$$

Finally because of (**) we get

$$\alpha_1-\frac{\tau_{N,1}}{\sigma_{N,1}}<\frac{1}{2^{s_{N+1}-1}}=\frac{1}{s^{Nm2^{m s_N}-1}}<\frac{1}{\sigma_{N,1}^l}$$

for every fixed l and all sufficiently big N. So after a theorem of LIOUVILLE [42, Theorem 191; HARDY-WRIGHT] α_1 must be transcendental. Then $\alpha_1,\cdots,\alpha_m$ are $\mathbf{Q}$–linearly independent. *Q.E.D.*

Dolbeault cohomology of toroidal theta groups

Let X be a complex manifold of dimension n, Ω^m the sheaf of germs of holomorphic m-forms

$$\omega=\sum_{|I|=m}h_I dz_I\qquad(m\leq n),$$

where $\mathcal{O}=\Omega^0$ is the sheaf of germs of holomorphic functions, $\mathcal{E}^{m,p}$ be the sheaf of germs of $\mathcal{C}^\infty$ complex valued (m,p)-forms

$$\omega=\sum_{|I|=m,|J|=p}f_{I,J}dz_I\wedge d\bar{z}_J\qquad(0\leq m,p\leq n)$$

where $\mathcal{E}:=\mathcal{E}^{0,0}$ is the sheaf of germs of $\mathcal{C}^\infty$-functions and finally $\Omega^m(X)$ be the holomorphic m-forms and $\mathcal{E}^{m,p}(X)$ the $\mathcal{C}^\infty$ (m,p)-forms on X. Then by Dolbeault's lemma (also called $\bar\partial$–Poincaré's lemma) there exists an exact sequence of fine sheaves

$$(\mathcal{E})\qquad 0\to\Omega^m\xrightarrow{\iota}\mathcal{E}^{m,0}\xrightarrow{\bar\partial}\mathcal{E}^{m,1}\xrightarrow{\bar\partial}\mathcal{E}^{m,2}\xrightarrow{\bar\partial}\cdots\xrightarrow{\bar\partial}\mathcal{E}^{m,n}\to 0,$$

which induces the DOLBEAULT sequence

$$0\to\Omega^m(X)\xrightarrow{\iota}\mathcal{E}^{m,0}(X)\xrightarrow{\bar\partial}\mathcal{E}^{m,1}(X)\xrightarrow{\bar\partial}\mathcal{E}^{m,2}(X)\xrightarrow{\bar\partial}\cdots\xrightarrow{\bar\partial}\mathcal{E}^{m,n}(X)\to 0$$

of global sections. This sequence defines the DOLBEAULT $\bar\partial$-cohomology groups

$$H^{m,p}(X):=Z(X,\mathcal{E}^{m,p})/B(X,\mathcal{E}^{m,p})$$

with the space $Z(X,\mathcal{E}^{m,p})=\ker(\mathcal{E}^{m,p}(X)\xrightarrow{\bar\partial}\mathcal{E}^{m,p+1}(X))$ of $\bar\partial$-closed and the space $B(X,\mathcal{E}^{m,p})=\operatorname{im}(\mathcal{E}^{m,p-1}(X)\xrightarrow{\bar\partial}\mathcal{E}^{m,p}(X))$ of $\bar\partial$-exact $\mathcal{C}^\infty$ (m,p)–forms on X. Following a theorem of DOLBEAULT [36, p 204; GRAUERT-FRITZSCHE] there exists an isomorphism

$$H^{m,p}(X) \simeq H^p(X, \Omega^m)$$

to Čech cohomology.

For groups $X = \mathbf{C}^n/\Lambda$ the lifts of (m,p)–forms on X are exactly the (m,p)–forms $\omega = \sum f_{J,K} dz_J \wedge d\bar{z}_K$ on $\mathbf{C}^n$ with Λ–periodic $\mathcal{C}^\infty$–coefficients $f_{J,K}$.

For toroidal groups $X = \mathbf{C}^n/\Lambda$ the set $\{dz_J : |J| = m\}$ is a basis of $H^{m,0}(X)$ $(0 \leq m \leq n)$.
For that let $\omega = \sum_{|J|=m} f_J dz_J$ be a $\bar{\partial}$–closed Λ–periodic $(0,m)$–form on $\mathbf{C}^n$. Then the coefficients must be holomorphic and therefore constant. So for toroidal groups

$$\dim H^0(X, \Omega^m) = \binom{n}{m} \qquad (0 \leq m \leq n).$$

To calculate all cohomology groups we use the fact that toroidal groups are principal $\mathbf{C}^{*n-q}$–fibre bundles over a torus group. We have seen in 1.1.12 that every Lie group $X = \mathbf{C}^n/\Lambda$ of rank $n+q$ has a closed subgroup isomorphic to $\mathbf{C}^{*n-q}$ so that $T := X/\mathbf{C}^{*n-q}$ becomes a torus group.

Vogt determined in 1981–1983 the cohomology groups of toroidal theta groups. Kazama and Umeno calculated in 1982–1984 the cohomology of all toroidal groups. All authors used the same principal idea, namely to work with forms which are holomorphic in the fibre coordinates of the natural $\mathbf{C}^{*n-q}$–bundle. [117, 56]

In this sense Kazama and Umeno considered in 1992 [58] a locally trivial fibre space X over a paracompact complex manifold T of complex dimension q with fibres which are biholomorphic to a Stein manifold S. With the sheaf of germs $\mathcal{F}$ of $\mathcal{C}^\infty$-functions which are holomorphic along the fibres and the sheaf $\mathcal{F}^{m,p}$ of germs of $\mathcal{C}^\infty$ (m,p)-forms with coefficients in $\mathcal{F}$ they got an analogous resolution of Ω^m and representation of $H^p(X, \Omega^m)$.

In details we take

$$\Lambda = \begin{pmatrix} 0 \\ \mathbf{Z}^{n-q} \end{pmatrix} \oplus \Gamma \quad \text{with the basis } P = \begin{pmatrix} 0 & I_q & S \\ I_{n-q} & R_1 & R_2 \end{pmatrix} \quad \text{where } \det \operatorname{Im} S \neq 0$$

in toroidal coordinates $(z, w) = (z_1, \cdots, z_q, w_1, \cdots, w_{n-q})$ so that with $B := (I_q, S)$ the projection

$$\pi : X = \mathbf{C}^n/\Lambda \longrightarrow \mathrm{M}\mathbf{C}_\Lambda / B\mathbf{Z}^{2q} = T$$

induces a principal $\mathbf{C}^{*n-q}$–fibre bundle with the q–dimensional torus group T as base space. Moreover let $\zeta = (z, w)$ be all n variables and as before Ω^m the sheaf of germs of all holomorphic m-forms

$$\omega = \sum_{|I|=m} h_I d\zeta_I \qquad (m \leq n).$$

Now we define $\mathcal{F}$ as the sheaf of germs of C^∞-functions on X which are holomorphic along the fibres and $\mathcal{F}^{m,p}$ as the sheaf of germs of (m,p)-forms with respect to $\{dz_1, \cdots, dz_n, d\overline{z}_1, \cdots, d\overline{z}_q\}$ and coefficients in $\mathcal{F}$.

Let u be the first q and v the last $n-q$ complex variables in toroidal coordinates. We decompose the $\overline{\partial}$-operator into

$$\overline{\partial} = \overline{\partial}_u + \overline{\partial}_v.$$

By Dolbeault's Lemma applied on the base space we get as $(\mathcal{E})$

$$(\mathcal{F}) \qquad 0 \to \Omega^m \xrightarrow{\iota} \mathcal{F}^{m,0} \xrightarrow{\overline{\partial}_u} \mathcal{F}^{m,1} \xrightarrow{\overline{\partial}_u} \mathcal{F}^{m,2} \xrightarrow{\overline{\partial}_u} \cdots \xrightarrow{\overline{\partial}_u} \mathcal{F}^{m,q} \to 0.$$

Now we can calculate the cohomology groups with the following lemma of Dolbeault type.
For that let $Z(X, \mathcal{F}^{m,p}) := \ker(\mathcal{F}^{m,p}(X) \xrightarrow{\overline{\partial}_u} \mathcal{F}^{m,p+1}(X))$ be the space of all $\overline{\partial}_u$-closed and $B(X, \mathcal{F}^{m,p}) := \operatorname{im}(\mathcal{F}^{m,p-1}(X) \xrightarrow{\overline{\partial}_u} \mathcal{F}^{m,p}(X))$ of all $\overline{\partial}_u$-exact forms.

2.2.4 Lemma (KAZAMA-DOLBEAULT)
Let $X = \mathbf{C}^n/\Lambda$ be a toroidal group. Then

$$H^p(X, \Omega^m) \simeq Z(X, \mathcal{F}^{m,p})/B(X, \mathcal{F}^{m,p}).$$

Proof
Let $\mathcal{U} = \{U_\alpha\}$ be a finite open covering of the torus T with trivialization

$$\pi^{-1}(U_\alpha) \simeq U_\alpha \times \mathbf{C}^{*n-q},$$

to be identified.
Moreover let $\mathcal{E}^{m',p}$ be the sheaf of C^∞ (m',p)-forms on U_α and $\Omega_*^{m''}$ the sheaf of holomorphic m''-forms on $\mathbf{C}^{*n-q}$. Then

$$\mathcal{F}^{m,p}|_{U_\alpha \times \mathbf{C}^{*n-q}} \simeq \bigoplus_{m'+m''=m} \mathcal{E}^{m',p} \hat{\otimes} \Omega_*^{m''},$$

where $\hat{\otimes}$ denotes the topological tensor product.
By KÜNNETH's formula [49, KAUP] we get

$$\begin{aligned} H^k(U_\alpha \times \mathbf{C}^{*n-q}, \mathcal{F}^{m,p}) &\simeq H^k(U_\alpha \otimes \mathbf{C}^{*n-q}, \bigoplus_{m'+m''=m} \mathcal{E}^{m',p} \hat{\otimes} \Omega_*^{m''}) \\ &\simeq \bigoplus_{\substack{s+t=k \\ m'+m''=m}} H^s(U_\alpha, \mathcal{E}^{m',p}) \hat{\otimes} H^t(\mathbf{C}^{*n-q}, \Omega_*^{m''}) = 0 \ (k \geq 1). \end{aligned}$$

Then $\mathcal{V} = \{U_\alpha \times \mathbf{C}^{*n-q}\}$ is a LERAY covering for $\mathcal{F}^{m,p}$ on X.

Let $\{\varrho_\alpha\}$ be a partition of unity subordinate to the covering $\mathcal{U} = \{U_\alpha\}$ of T. For a cocycle $\{f_{\alpha_0\alpha_1\cdots\alpha_k}\} \in Z^k(\mathcal{V}, \mathcal{F}^{m,p})$ we put

$$g_{\alpha_0\alpha_1\cdots\alpha_{k-1}}(x) := \sum_\alpha \varrho_\alpha(\pi(x)) f_{\alpha\alpha_0\alpha_1\cdots\alpha_{k-1}}(x).$$

Then $\{g_{\alpha_0\alpha_1\cdots\alpha_{k-1}}\} \in C^{k-1}(\mathcal{V}, \mathcal{F}^{m,p})$ becomes a cochain and $\delta\{g_{\alpha_0\alpha_1\cdots\alpha_{k-1}}\} = \{f_{\alpha_0\alpha_1\cdots\alpha_k}\}$ because $\{f_{\alpha_0\alpha_1\cdots\alpha_k}\}$ is a cocycle. Hence $H^k(\mathcal{V}, \mathcal{F}^{m,p}) = 0$ $(k \geq 1)$. This and $(\mathcal{F})$ proves the theorem. *Q.E.D.*

For the next step we follow VOGT [117].

Two (m,p)–forms are $\overline{\partial}$*–cohomologous*, iff their difference is $\overline{\partial}$–exact.

We want to show that every $\overline{\partial}$–closed (m,p)–form on a toroidal theta group is $\overline{\partial}$–cohomologous to a form with constant coefficients. For that it is sufficient to consider $\overline{\partial}$–closed $(0,p)$–forms.
Indeed, with

$$\sum_{I,J} F_{IJ} dz_I \wedge d\overline{z}_J = \sum_I dz_I \wedge (\sum_J F_{IJ} d\overline{z}_J)$$

all $\sum_J F_{IJ} d\overline{z}_J$ are $\overline{\partial}$–closed.

Let u be the first q und v the last $n-q$ toroidal coordinates after *refined* transformation from standard coordinates so that Λ has the basis

$$P = \begin{pmatrix} 0 & B \\ I_{n-q} & R \end{pmatrix},$$

where $B = (B_1, B_2)$ with $B_1 = (\mathrm{Im}\hat{T})^{-1}$ and $\mathrm{Re}B_2 = (\mathrm{Im}\hat{T})^{-1}\mathrm{Re}\hat{T}$ is the basis of a torus T and $R = (R_1, R_2)$ the glueing matrix. We get a parametrization $L_{\mathbf{R}}(t) = At \quad (t \in \mathbf{R}^{2n})$ of the *real* coordinates (Reu, Imu, Rev, Imv) according 1.1.13 $(L_{\mathbf{R}})$, where the inverse matrix is

$$(*) \qquad A^{-1} = \begin{pmatrix} \mathrm{Im}\hat{T} & -\mathrm{Re}\hat{T} & 0 & 0 \\ 0 & I_q & 0 & 0 \\ -R_1\mathrm{Im}\hat{T} & R_1\mathrm{Re}\hat{T} - R_2 & I_{n-q} & 0 \\ 0 & 0 & 0 & I_{n-q} \end{pmatrix}.$$

2.2.5 Proposition (VOGT)
Let $\mathbf{C}^n/\Lambda$ be a toroidal *theta* group. Then every $\overline{\partial}$–closed Λ–periodic $(0,p)$–form is $\overline{\partial}$–cohomologous to a form with constant coefficients.
Proof
Let $\omega = \sum_{|J|=p} F_J d\overline{z}_J$ be Λ–periodic.
a) The coefficients $F_J \circ L_{\mathbf{R}}^{-1}$ are $\mathbf{Z}^{n+q}$–periodic in the first $n+q$ variables t' of

the real parameters $t = (t', t'') \in \mathbf{R}^{2n}$. So we can develop F_J into a real Fourier series

$$F_J \circ L_{\mathbf{R}}^{-1}(t) = \sum_{\sigma \in \mathbf{Z}^{n+q}} f_J^{*(\sigma)}(t'')\, \mathbf{e}(\langle \sigma, t' \rangle).$$

We get $F_J(u,v) = \sum_{\sigma \in \mathbf{Z}^{n+q}} f_J^{*(\sigma)}(\mathrm{Im}v)\, \mathbf{e}[E^{(\sigma)}(u,v) - \langle \sigma_3, i\mathrm{Im}v \rangle]$ with the help of $(*)$, where

$E^{(\sigma)}(u,v) :=$

$$\left[({}^t\sigma_1 - {}^t\sigma_3 R_1)\mathrm{Im}\hat{T}\right] \mathrm{Re}\, u - \left[({}^t\sigma_1 - {}^t\sigma_3 R_1)\mathrm{Re}\hat{T} - ({}^t\sigma_2 - {}^t\sigma_3 R_2)\right] \mathrm{Im}\, u + \langle \sigma_3, v \rangle,$$

and $\sigma_1, \sigma_2 \in \mathbf{Z}^q$ and $\sigma_3 \in \mathbf{Z}^{n-q}$ are the components of $\sigma \in \mathbf{Z}^{n+q}$.

The previous Lemma shows that the F_J can be assumed to be holomorphic in the last $n-q$ complex variables v. But then the $f_J^{(\sigma)} := f_J^{*(\sigma)}(\mathrm{Im}v)\, \mathbf{e}(-\langle \sigma_3, i\mathrm{Im}v \rangle)$ must be constant so that

$$F_J = \sum_{\sigma \in \mathbf{Z}^{n+q}} f_J^{(\sigma)}\, \mathbf{e}(E^{(\sigma)})$$

with constant coefficients.

Now define

$$\omega^{(\sigma)} := \sum_{|J|=p} f_J^{(\sigma)} \mathbf{e}(E^{(\sigma)}) d\bar{z}_J,$$

$$\vartheta^{(\sigma)} := \sum_{|J|=p} f_J^{(\sigma)} d\overline{z_J},$$

$$C^{(\sigma)} := \frac{1}{2}\left(({}^t\sigma_1 - {}^t\sigma_3 R_1)\mathrm{Im}\hat{T} - i\left[({}^t\sigma_1 - {}^t\sigma_3 R_1)\mathrm{Re}\hat{T} - ({}^t\sigma_2 - {}^t\sigma_3 R_2)\right]\right)$$

and

$$\xi^{(\sigma)} := \sum_{j=1}^{q} C_j^{(\sigma)} d\bar{z}_j.$$

Then

$$\bar{\partial}\omega^{(\sigma)} = 2\pi i \mathbf{e}(E^{(\sigma)})\xi^{(\sigma)} \wedge \vartheta^{(\sigma)}.$$

Since $\omega = \sum_{\sigma \in \mathbf{Z}^{n+q}} \omega^{(\sigma)}$ is $\bar{\partial}$–closed and $\xi^{(0)} = 0$,

$$(\dagger) \qquad \bar{\partial}\omega = 2\pi i \sum_{\sigma \neq 0} \mathbf{e}(E^{(\sigma)})\xi^{(\sigma)} \wedge \vartheta^{(\sigma)} = 0.$$

It is to remark that $C^{(\sigma)} \neq 0$ for $\sigma \neq 0$. Otherwise ${}^t\sigma_1 = {}^t\sigma_3 R_1$ and ${}^t\sigma_2 = {}^t\sigma_3 R_2$ would contradict the irrationality condition 1.1.12(I) for toroidal groups.

b) It is well known [77, p 7; MUMFORD] that for any $\mathbf{C}$–vectorspace V and any $\mathbf{C}$–linear map $D : V \to \mathbf{C}$ there exists a map

$$D\rfloor : \bigwedge^{m} V \to \bigwedge^{m-1} V$$

with the properties:

$$(*) \quad D\rfloor(X_1 \wedge \cdots \wedge X_m) = \sum_{k=1}^{m} (-1)^{m-k} D(X_k) X_1 \wedge \cdots \wedge \hat{X}_k \wedge \cdots \wedge X_m$$

and

(**) If $D(X_0) = 1$, then for every $\alpha \in \bigwedge^l V$

$$D\rfloor(\alpha \wedge X_0) + (D\rfloor\alpha) \wedge X_0 = \alpha.$$

$D\rfloor$ is the so-called *interior multiplication* by D. Now define the $\mathbf{C}$–linear space

$$V := \left\{ \sum_{j=1}^{q} a_j d\bar{z}_j : \ a_j \in \mathbf{C} \quad (j = 1, \cdots, q) \right\},$$

the $\mathbf{C}$–linear map $D^{(\sigma)} : V \to \mathbf{C}$ by

$$D^{(\sigma)}(\sum_{j=1}^{q} a_j d\bar{z}_j) := \sum_{j=1}^{q} a_j \frac{\overline{C_j^{(\sigma)}}}{\left|C^{(\sigma)}\right|^2} \qquad (\sigma \in \mathbf{Z}^{n+q} \setminus \{0\})$$

and

$$\eta^{(\sigma)} := \frac{(-1)^{p-1}}{2\pi i} \mathbf{e}(E^{(\sigma)}) D^{(\sigma)}\rfloor(\vartheta^{(\sigma)}) \qquad (\sigma \in \mathbf{Z}^{n+q} \setminus \{0\}).$$

Then

$$\overline{\partial}_{\eta}^{(\sigma)} = \mathbf{e}(E^{(\sigma)}) \left(D^{(\sigma)}\rfloor(\vartheta^{(\sigma)}) \right) \wedge \xi^{(\sigma)}.$$

Because $D^{(\sigma)}(\xi^{(\sigma)}) = 1$ we get with (**)

$$D^{(\sigma)}\rfloor(\vartheta^{(\sigma)} \wedge \xi^{(\sigma)}) + \left(D^{(\sigma)}\rfloor(\vartheta^{(\sigma)}) \right) \wedge \xi^{(\sigma)} = \vartheta^{(\sigma)}$$

so that

$$\mathbf{e}(E^{(\sigma)}) D^{(\sigma)}\rfloor(\vartheta^{(\sigma)} \wedge \xi^{(\sigma)}) + \overline{\partial}\eta^{(\sigma)} = \omega^{(\sigma)} \qquad (\sigma \in \mathbf{Z}^{n+q} \setminus \{0\})$$

and with (†) $\sum_{\sigma \neq 0} \overline{\partial}\eta^{(\sigma)} = \sum_{\sigma \neq 0} \omega^{(\sigma)} = \omega - \omega^{(0)}$.

If we define *formally* $\eta := \sum_{\sigma \neq 0} \eta^{(\sigma)}$, then ω becomes $\overline{\partial}$–cohomologous to $\omega^{(0)} = \sum_{|J|=p} f_J^{(0)} d\bar{z}_J$, a form with constant coefficients. But we can show that $\sum_{\sigma \neq 0} \eta^{(\sigma)}$ is convergent for toroidal theta groups only.

Remember that every coefficient of $\eta^{(\sigma)}$ is a finite sum of summands

$$\frac{(-1)^{p-1}}{2\pi i} \cdot \frac{C_j^{(\sigma)}}{\left|C^{(\sigma)}\right|^2} f_J^{(\sigma)} \mathrm{e}(E^{(\sigma)}).$$

The number of the summands is independent of σ and $\sum_\sigma \left|f_J^{(\sigma)}\right| k^{\sigma_3}$ convergent for every $k \in \mathbf{R}^q_{>0}$ because the coefficients F_J are holomorphic in v.

Moreover remember (TT) of Theorem 2.2.2 so that for a suitable real $r > 0$

$$r^{-|\sigma_3|} \leq \inf_{\tau \in \mathbf{Z}^{2q}} \left|{}^t\tau - {}^t\sigma_3 R\right| \qquad (\sigma_3 \in \mathbf{Z}^{n-q} \setminus \{0\}).$$

With $\tau := \binom{\sigma_1}{\sigma_2} \in \mathbf{Z}^{2q}$ we have

$${}^tC^{(\sigma)} = \frac{1}{2}({}^t\tau - {}^t\sigma_3 R) \begin{pmatrix} \mathrm{Im}\hat{T} & -i\mathrm{Re}\hat{T} \\ 0 & iI_q \end{pmatrix}$$

so that

$$r^{-|\sigma_3|} \leq d\left|C^{(\sigma)}\right| \qquad (\sigma \in \mathbf{Z}^{n+q} \text{ with } \sigma_3 \neq 0)$$

with a real constant $d > 0$. Then $\sum_{\sigma \neq 0} \frac{\left|f_J^{(\sigma)}\right|}{\left|C^{(\sigma)}\right|} k^{\sigma_3} \leq d \sum_{\sigma \neq 0} \left|f_J^{(\sigma)}\right| r^{|\sigma_3|} k^{\sigma_3}$ is convergent for every $k \in \mathbf{R}^q_{>0}$, even in the case of $\sigma_3 = 0$. *Q.E.D.*

We have seen that every Λ–periodic $(0,p)$–form on a toroidal theta group is $\overline{\partial}$-cohomologous to a form $\sum_{|J|=p} f_J d\overline{z}_J$ with constant coefficients $f_J \in \mathbf{C}$. In addition to that we can assume that in toroidal coordinates every $f_J = 0$, if there exists a $j \in J$ with $j > q$.

Indeed, for every $j > q$ the function $f_j := \overline{z_j} - z_j$ is Λ–periodic and $\overline{\partial} f_j = d\overline{z}_j$ so that all $d\overline{z}_j \wedge d\overline{z}_J = \overline{\partial}(f_j d\overline{z}_J)$ are $\overline{\partial}$–exact.

On toroidal theta groups every element of $H^{m,p}(X)$ is represented by a ***translation invariant harmonic form*** which is ***not uniquely determined for*** $q < n$.

Now we get the final result about toroidal theta groups

2.2.6 Theorem

Let $X = \mathbf{C}^n/\Lambda$ be a toroidal *theta* group of type q. Then:

1. If $X \xrightarrow{\pi} T$ represents X as $\mathbf{C}^{*n-q}$–fibre bundle over the torus T, then π induces an isomorphism

$$\pi^* : H^p(T, \mathcal{O}) \to H^p(X, \mathcal{O}).$$

2. a) Every differential form in $Z(X, \mathcal{E}^{m,p})$ is represented by a uniquely determined

$$\frac{1}{m!}\frac{1}{p!}\sum_{\substack{|J|=m\\|K|=p}} c_{JK}\,dz_J\wedge d\overline{z}_K \in \bigwedge^m \mathbf{C}\{dz_1,\cdots,dz_n\}\wedge\bigwedge^p \mathbf{C}\{d\overline{z}_1,\cdots,d\overline{z}_q\}.$$

b)

$$H^p(X,\Omega^m)\simeq\bigwedge^m \mathbf{C}\{dz_1,\cdots,dz_n\}\wedge\bigwedge^p \mathbf{C}\{d\overline{z}_1,\cdots,d\overline{z}_q\},$$

then

$$\dim H^p(X,\Omega^m)=\binom{n}{m}\binom{q}{p}.$$

Proof

1. With Dolbeault isomorphism and in toroidal coordinates we can restrict our considerations to $(0,p)$–forms with constant coefficients $\sum_{|J|=p} c_J d\overline{z}_J$ and $j\le q$ for all $j\in J$. Their forms are forms on T so that the lift from T to X is also surjective and injective.

2. $H^p(X,\Omega^m)\simeq H^{m,p}(X)$ is represented in toroidal coordinates after Proposition 2.2.5 and the previous remarks by all (m,p)–forms $\sum_{\substack{|I|=m\\|J|=p}} c_{I,J}dz_I\wedge d\overline{z}_J$ with constant coefficients where all $j\le q$ for all $j\in J$ in the sum.

We have the decomposition

$$\omega=\sum_I dz_I\wedge(\sum_J c_{I,J}d\overline{z}_J)=\sum_I(dz_I\wedge\omega_I).$$

ω is $\overline{\partial}$–exact, iff the ω_I are so for all I. So we get the result by 1). *Q.E.D.*

Dolbeault cohomology of complex Lie groups

MALGRANGE took in 1975 an example of GRAUERT of a pseudoconvex complex manifold with only constant holomorphic functions and proved, that its cohomology group $H^1(X,\mathcal{O})$ has a quasi–Fréchet non–Hausdorff topology [67, 84]. KAZAMA determined in 1984 the cohomology groups of all toroidal goups [53]. Their result about toroidal wild groups is the

2.2.7 Proposition

Let X be a toroidal wild group of type q. Then for $1\le p\le q$ the $H^p(X,\mathcal{O})$ are infinite dimensional vector spaces with non–Hausdorff topology.

Proof

According to (TT) of Theorem 2.2.2(4) a toroidal group is wild, iff in toroidal coordinates

$$(W)\quad \forall N\in\mathbf{N}_{>0}\,\exists\sigma_{3,N}\in\mathbf{Z}^{n-q}\setminus\{0\},\tau_N\in\mathbf{Z}^{2q}:\left|{}^tR\sigma_{3,N}-\tau_N\right|\le N^{-|\sigma_{3,N}|}$$

with glueing matrix R. Denote with $\sigma_{1,N}, \sigma_{2,N} \in \mathbf{Z}^q$ the components of τ_N and ${}^t\sigma := ({}^t\sigma_1, {}^t\sigma_2, {}^t\sigma_3) \in \mathbf{Z}^{n+q}$. Let us remember the proof of Proposition 2.2.5. We had seen that $\sum \overline{\partial}\eta^{(\sigma)}$ with

$$\eta^{(\sigma)} := \frac{(-1)^{p-1}}{2\pi i} \mathrm{e}\left(E^{(\sigma)}\right) D^{(\sigma)} \rfloor (\vartheta^{(\sigma)})$$

is convergent so that especially $\sum \overline{\partial}_\eta^{(\sigma_N)}$ is convergent. It is easy to explain that $\sum \eta^{(\sigma_N)}$ can be divergent in the case of toroidal *wild* groups.
Indeed, in the same way as in that proof we get by (W)

$$d\left|C^{(\sigma_N)}\right| \le N^{-|\sigma_{3,N}|}$$

so that for a certain j_0

$$\sum \frac{\left|C_{j_0}^{(\sigma_N)}\right|}{\left|C^{(\sigma_N)}\right|^2} \left|f_J^{(\sigma_N)}\right| k^{\sigma_{3,N}} \ge \frac{1}{\sqrt{q}} \sum \frac{\left|f_J^{(\sigma_N)}\right|}{\left|C^{(\sigma_N)}\right|} k^{\sigma_{3,N}} \ge \frac{d}{\sqrt{q}} \left|f_J^{(\sigma_N)}\right| N^{|\sigma_{3,N}|} k^{\sigma_{3,N}}.$$

Take $\left|f_J^{(\sigma_N)}\right| = N^{-|\sigma_N|}$ and $f_J^{(\sigma)} = 0$ for the other σ. Then $\sum_N f_J^{(\sigma_N)} k^{\sigma_{3,N}}$ is convergent for every $k \in \mathbf{R}^{n-q}_{>0}$ but $\sum \eta^{(\sigma_N)}$ divergent.
Decisive is the
Assertion. $\tilde{\omega} := \sum_N \overline{\partial}\eta^{(\sigma_N)}$ is not $\overline{\partial}$–exact. This proved the proposition is proved, because then $\tilde{\omega}$ is in the closure of the space of the $\overline{\partial}$–exact $(0,p)$–forms but not in this space itself so that $H^{0,p}(X)$ cannot be Hausdorff.
Proof of the assertion.
Assume that $\tilde{\omega} = \overline{\partial}\gamma$ with $\gamma = \sum_{|J|=p-1} G_J d\overline{z}_J$ and $G_J = \sum_\sigma g_J^{(\sigma)} \mathrm{e}(E^{(\sigma)})$ as in the proof of 2.2.5. By uniqueness of these decompositions we get

$$\overline{\partial}\eta^{(\sigma_N)} = \overline{\partial}\gamma^{(\sigma_N)}$$

with divergent $\sum_N \eta^{(\sigma_N)}$ but convergent $\sum_N \gamma^{(\sigma_N)}$, if we take $\eta^{(\sigma_N)}$ as above. We follow the proof of Proposition 2.2.5 and get

$$D^{(\sigma_N)} \rfloor (\vartheta_f^{(\sigma_N)}) \wedge \xi^{(\sigma_N)} = 2\pi i (-1)^{p-1} \vartheta_g^{(\sigma_N)} \wedge \xi^{(\sigma_N)}$$

with

$$\vartheta_f^{(\sigma_N)} = \sum_{|J|=p} f_J^{(\sigma_N)} d\overline{z}_J,\ \vartheta_g^{(\sigma_N)} = \sum g_J^{(\sigma_N)} d\overline{z}_J.$$

We want to compare the coefficients of $d\overline{z}_J = d\overline{z}_1 \wedge \cdots \wedge d\overline{z}_p$ only and get by (*) of the proof of Proposition 2.2.5

$$\sum_{k=1}^{p} f_J^{(\sigma_N)} C_k^{(\sigma_N)} D^{(\sigma_N)} \rfloor (d\overline{z}_k) = 2\pi i \sum_{k=1}^{p} (-1)^{k-1} g_{J_k}^{(\sigma_N)} C_k^{(\sigma_N)},$$

where the sum of the right side is taken over all $\hat{J}_k = (1, \cdots, \hat{k}, \cdots, p)$. Then

$$f_J^{(\sigma_N)} = 2\pi i \sum_{k=1}^{p} (-1)^{k-1} g_{\hat{J}_k}^{(\sigma_N)} C_k^{(\sigma_N)}.$$

Because $\sum g_{\hat{J}_k}^{(\sigma_N)} k^{\sigma_{3,N}}$ is convergent, $\sum \frac{f_J^{(\sigma_N)}}{|C^{(\sigma_N)}|} k^{\sigma_{3,N}}$ must be convergent for every $k \in \mathbf{R}_{>0}^{n-q}$. But this is not possible for the special sequence (σ_N) as we have seen above. *Q.E.D.*

The results in Theorem 2.2.6 are also sufficient for the characterization of toroidal theta groups. So KAZAMA calls them *toroidal groups of cohomologically finite type* and corresponding the toroidal wild groups *toroidal groups of cohomologically infinite type.*

For the case of general complex Lie groups we define:

2.2.8 Definition
Let X be a connected complex Lie group with maximal toroidal subgroup X_0.
X is a **Lie theta group**, iff X_0 is a toroidal theta group,
X is a **Lie wild group**, iff X_0 is a toroidal wild group.

For Lie theta and wild groups KAZAMA and UMENO got in 1990 the following result by studying their maximal toroidal subgroup [57]:

2.2.9 Theorem (KAZAMA–UMENO)
Let X be a connected complex Lie group of type q. Then:

1. X is a Stein group, iff all $H^p(X, \mathcal{O}) = 0$ $(p > 0)$.
   In the following cases let p be with $1 \leq p \leq q$:
2. X is a toroidal theta group, iff all $H^p(X, \mathcal{O})$ have a positive finite dimension.
3. X is a Lie theta group but not a toroidal theta group, iff all $H^p(X, \mathcal{O})$ are infinite dimensional with Hausdorff topology.
   With the maximal toroidal subgroup $X_0 \neq X$ of X

$$H^p(X, \mathcal{O}) \simeq H^0(X/X_0, \mathcal{O}) \otimes H^p(X_0, \mathcal{O}).$$

4. X is a Lie wild group, iff all $H^p(X, \mathcal{O})$ are infinite dimensional with non–Hausdorff topology.

For the *proof* we quote the original paper.

3. Quasi–Abelian Varieties

Quasi–Abelian varieties are toroidal groups with positive line bundles. KOPFERMANN *used in 1964 the Hermitian decomposition of automorphic factors to establish the period relations for toroidal groups.* GHERARDELLI *and* ANDREOTTI *contributed in 1971-73 the fibration theorem using ample Riemann forms.* ABE *characterized in 1987-89 quasi–Abelian varietes from different standpoints. The consequence of his theory is the Main Theorem and the meromorphic reduction of toroidal groups.* CAPOCASA *and* CATANESE *added in 1991 the existence of a non–degenerate meromorphic function.*

3.1 Ample Riemann forms

The Hermitian decomposition and the Appell–Humbert decomposition of an automorphic factor determine a Hermitian form whose properties can be described by the period relations. Ample Riemann forms define quasi–Abelian varieties which can be characterized in a first step by the fibration theorem.

The Hermitian decomposition of an automorphic factor

The Hermitian decomposition of theta factors was introduced by KOPFERMANN basing on the characteristic decomposition of $\mathbf{Z}^n$-periodic theta factors. Considerations of A. WEIL were used for this generalization. [64, 119]

3.1.1 Theorem (The Hermitian decomposition of an automorphic factor) (KOPFERMANN)
Let $\Lambda = \mathbf{Z}^n \oplus \Gamma \subset \mathbf{C}^n$ be a lattice of rank $n+q$. Then every $\mathbf{Z}^n$-periodic automorphic factor α_λ $(\lambda \in \Lambda)$ has the **Hermitian decomposition**

$$\alpha_\lambda(z) = \varrho(\lambda)\mathbf{e}\Big(\frac{1}{2i}\Big[(H+S)(z,\lambda) + \frac{1}{2}(H+S)(\lambda,\lambda)\Big] + s_\lambda(z) + h(\lambda)\Big) \qquad (\lambda \in \Lambda)$$

with the following properties:

1. H is a Hermitian form on $\mathbf{C}^n$ such that $\operatorname{Im} H|_{\Lambda\times\Lambda} = A$ is the characteristic bilinear and alternating $A : \Lambda \times \Lambda \to \mathbf{C}$,
   S is a symmetric $\mathbf{C}$-bilinear form on $\mathbf{C}^n$.

2. s_λ $(\lambda \in \Lambda)$ is the $\mathbf{Z}^n$-periodic wild automorphic summand with vanishing 0-coefficients.
3. The map $\varrho : \Lambda \to \mathbf{S}^1 := \{z : |z| = 1\}$ is a *semi-character* of Λ, hence

$$\varrho(\lambda + \lambda') = \varrho(\lambda)\varrho(\lambda')\mathbf{e}\Big(\frac{1}{2}A(\lambda, \lambda')\Big) \qquad (\lambda, \lambda' \in \Lambda),$$

and $h : \mathbf{C}^n \to \mathbf{C}$ a $\mathbf{C}$–linear form.

Uniqueness and cobordism.
1. The $\mathbf{C}$–linear form

$$L(z, \lambda) := \frac{1}{2i}(H + S)(z, \lambda) = \langle \chi_\lambda, z \rangle \quad (z \in \mathbf{C})$$

is the uniquely determined characteristic homomorphism $\langle \chi_\lambda, z \rangle$ which depends only on the line bundle defined by the given automorphic factor α_λ $(\lambda \in \Lambda)$. The ImH is uniquely determined on $\mathbf{R}_\Lambda$.
2. The wild summand $s_\lambda(z)$ is uniquely determined by the given automorphic factor α_λ $(\lambda \in \Lambda)$.
For cobordant factors with the same properties as in the theorem the wild summands are strictly cobordant with $\mathbf{Z}^n$-periodic cobordism functions.
3. The product $\varrho(\lambda)\mathbf{e}(h(\lambda))$ is the uniquely determined constant part in the decomposition of α_λ $(\lambda \in \Lambda)$. The decomposition into a semi–character ϱ and a linear form h is unique only on the maximal $\mathbf{C}$–linear subspace $\mathbf{MC}_\Lambda$ of $\mathbf{R}_\Lambda$.
Proof
i) Definition of the Hermitian form H with Im $H|_{\Lambda \times \Lambda} = A$.
Let $L(z, \lambda) := \langle \chi_\lambda, z \rangle$ $(z \in \mathbf{C}^n)$ be the characteristic linear form of the line bundle L defined by the given automorphic factor as in the characteristic decomposition 2.1.3. Then every $L(z, \lambda)$ $(\lambda \in \Lambda)$ has a unique $\mathbf{R}$–linear extension

$$L(z, v) \qquad (v \in \mathbf{R}_\Lambda)$$

and then

$$A(u, v) = L(v, u) - L(u, v) \qquad (u, v \in \mathbf{R}_\Lambda)$$

is the unique real-valued and alternating $\mathbf{R}$-bilinear extension of the characteristic bilinear $A : \Lambda \times \Lambda \to \mathbf{Z}$ to $\mathbf{R}_\Lambda \times \mathbf{R}_\Lambda$.
Remember that for a real-valued $\mathbf{R}$-bilinear form R and a real-valued and alternating $\mathbf{R}$-bilinear form A the condition

$$R(u, v) = A(iu, v) = -A(u, iv)$$

holds, iff $H := R + iA$ is Hermitian.
But Im$L(u, v)$ is symmetric because $A(u, v)$ is real and Im$L(iu, v)$ =Re$L(u, v)$ because $L(u, v)$ is $\mathbf{C}$–linear in u. Therefore $A(iu, v) + A(u, iv) = 0$ for all $u, v \in \mathbf{MC}_\Lambda$. Hence, if we define $R(u, v) := A(iu, v)$ $(u \in \mathbf{MC}_\Lambda, v \in \mathbf{R}_\Lambda)$,

$$H(u,v) := R(u,v) + iA(u,v) \qquad (u \in \mathbf{MC}_\Lambda, v \in \mathbf{R}_\Lambda)$$

is Hermitian on $\mathbf{MC}_\Lambda \times \mathbf{MC}_\Lambda$, the unique one with $\mathrm{Im}H = A$ on this subspace. To extend H put $\mathbf{R}_\Lambda = \mathbf{MC}_\Lambda \oplus V$ with an $\mathbf{R}$-vectorspace V and take *any* real valued and symmetric $\mathbf{R}$-bilinear form R_V on $V \times V$. Then define the $\mathbf{R}$-valued and symmetric $\mathbf{R}$-bilinear extension R on $\mathbf{R}_\Lambda \times \mathbf{R}_\Lambda$ by given R on $(\mathbf{MC}_\Lambda \times \mathbf{R}_\Lambda) \cup (\mathbf{R}_\Lambda \times \mathbf{MC}_\Lambda)$ and by R_V on $V \times V$. Then

$$H(u,v) := R(u,v) + iA(u,v) \qquad (u, v \in \mathbf{R}_\Lambda)$$

has after fixing any R_V a unique Hermitian extension to $\mathbf{C}^n \times \mathbf{C}^n$.

ii) The decomposition of the automorphic factor.
1) Obviously

$$S(z,v) := 2iL(z,v) - H(z,v) \qquad (z \in \mathbf{C}^n, v \in \mathbf{R}_\Lambda)$$

is $\mathbf{C}$–linear in z and symmetric on $\mathbf{R}_\Lambda \times \mathbf{R}_\Lambda$. Therefore there exists a unique symmetric extension of S to $\mathbf{C}^n \times \mathbf{C}^n$. Then

$$L(z,\lambda) = \langle \chi_\lambda, z \rangle = \frac{1}{2i}(H+S)(z,\lambda)$$

is the uniquely determined characteristic linear form of the characeristic decomposition 2.1.3.
2) s_λ $(\lambda \in \Lambda)$ is the wild summand of the characeristic decomposition 2.1.3.
3) The real homomorphism $d : \Lambda \to \mathbf{R}$ has a unique $\mathbf{R}$–linear extension $d : \mathbf{R}_\Lambda \to \mathbf{R}$. Now

$$h(u) := d(iu) + id(u) \qquad (u \in \mathbf{R}_\Lambda)$$

is the unique $\mathbf{C}$–linear form on $\mathbf{MC}_\Lambda$ with $\mathrm{Im}h = d$. Moreover let $r : \mathbf{R}_\Lambda \to \mathbf{R}$ be *any* $\mathbf{R}$–linear extension of $\mathrm{Re}h$ from $\mathbf{MC}_\Lambda$ to $\mathbf{R}_\Lambda$. Then

$$h(u) := r(u) + id(u) \qquad (u \in \mathbf{R}_\Lambda)$$

has after fixing r a unique $\mathbf{C}$–linear extension to $\mathbf{C}^n$.
Finally

$$\varrho(\lambda) := \mathbf{e}(c_\lambda - r(\lambda)) \qquad (\lambda \in \Lambda)$$

with c_λ as in the characteristic decomposition 2.1.3 is a semi–character for A.

iii) The rest is clear. *Q.E.D.*

Of course for lattices $\Lambda = \mathbf{Z}^n \oplus \Gamma \subset \mathbf{C}^n$ the Hermitian decomposition can be used in Γ–reduced form.

Now it is very easy to get the following decomposition well known in classical torus theory:

3.1.2 Corollary (Appell–Humbert decomposition)
Let $X = \mathbf{C}^n/\Lambda$ be an Abelian Lie group with $\Lambda = \mathbf{Z}^n \oplus \Gamma$.
Then every line bundle L on X can be given by an automorphic factor

$$\beta_\lambda(z) := \varrho(\lambda)\mathbf{e}\Big(\frac{1}{2i}\Big[H(z,\lambda) + \frac{1}{2}H(\lambda,\lambda)\Big] + s_\lambda(z)\Big) \qquad (\lambda \in \Lambda)$$

with Hermitian H, wild summand s_λ and semi–character ϱ as in the Hermitian decomposition 3.1.1.

Proof
Define the quadratic polynomial

$$q(z) := \frac{1}{4i}S(z,z) + h(z).$$

Then $q(z+\lambda) - q(z) = \frac{1}{4i}[S(z,\lambda) + \frac{1}{2}S(\lambda,\lambda)] + h(\lambda)$ so that β_λ becomes cobordant to α_λ $(\lambda \in \Lambda)$. *Q.E.D.*

The APPELL–HUMBERT decomposition cannot be used in Γ–reduced form, because the automorphic factor is nomore $\mathbf{Z}^n$-periodic. Informations about the characteristic homomorphism go lost.

The lost part of the given automorphic factor in the Appell–Humbert decomposition

$$\mathbf{e}\Big(\frac{1}{2i}\Big[S(z,\lambda) + \frac{1}{2}S(\lambda,\lambda)\Big] + h(\lambda)\Big) \qquad (\lambda \in \Lambda)$$

is obviously cobordant to $\mathbf{1}$. It defines the analytically trivial bundle $X \times \mathbf{C}$ and is therefore called a ***trivial theta factor.***

Every automorphic factor is the product of a theta factor, defined by its type, and a wild factor:

3.1.3 Definition
Let $X = \mathbf{C}^n/\Lambda$ be a group with Λ of rank $n+q$ and L be theta bundle on X with the characteristic alternating and bilinear form $A : \Lambda \times \Lambda \to \mathbf{Z}$ defined by the Chern class $c_1(L)$.

A **theta factor of type** (H, ϱ, S, h) for A is an automorphic factor

$$\vartheta_\lambda(z) = \varrho(\lambda)\mathbf{e}\Big(\frac{1}{2i}\Big[(H+S)(z,\lambda) + \frac{1}{2}(H+S)(\lambda,\lambda)\Big] + h(\lambda)\Big) \qquad (\lambda \in \Lambda)$$

where

1. H is a Hermitian with $\mathrm{Im}H|_{\Lambda\times\Lambda} = A$,
2. ϱ a semi–character for A,

3. S is a symmetric **C**-bilinear form and
4. h a **C**–linear form on $\mathbf{C}^n$.

A **reduced theta factor** is an automorphic factor

$$\vartheta_\lambda(z) = \varrho(\lambda)\mathbf{e}\Big(\frac{1}{2i}\Big[H(z,\lambda) + \frac{1}{2}H(\lambda,\lambda)\Big]\Big) \qquad (\lambda \in \Lambda)$$

of type $(H, \varrho, 0, 0)$ with H and ϱ as before. For simplicity we write (H, ϱ) instead of $(H, \varrho, 0, 0)$.

A **wild factor**

$$\iota_\lambda(z) = \mathbf{e}(s_\lambda(z)) \qquad (\lambda \in \Lambda)$$

is the exponential of of an automorphic summand s_λ $(\lambda \in \Lambda)$.

Together with Proposition 2.1.6 and the Decomposition Theorem 2.1.10 of Vogt we get the

3.1.4 Theorem (Decomposition of line bundles)
Every line bundle on a toroidal group $X = \mathbf{C}^n/\Lambda$ is the product $L = L_\vartheta \otimes L_0$ of a theta bundle L_ϑ and a topologically trivial line bundle L_0.
L_ϑ is defined by a theta factor of type (H, ϱ, S, h) as well by a theta factor of reduced type (H, ϱ) and L_0 by an wild factor $\iota_\lambda = \mathbf{e}(s_\lambda)$ $(\lambda \in \Lambda)$.
The wild factor ι_λ can be assumed to be constant on MC_Λ.

Every element of the Néron–Severi group NS(X) *can be represented by a theta bundle which is defined by a reduced theta factor.*

Period relations

We restrict our considerations to line bundles defined by reduced theta factors. Let $\Lambda \subset \mathbf{C}^n$ be a lattice of rank $r := n + q$ and ϑ_λ $(\lambda \in \Lambda)$ be a reduced theta factor which defines a line bundle L on $X = \mathbf{C}^n/\Lambda$ with Chern class $c_1(L)$ given by the characteristic alternating form A on Λ. The characteristic homomorphism $\chi : \Lambda \to \mathbf{C}^n$ can have complex values according to the characeristic decomposition 2.1.3, if we don't restrict to lattices in special situation. Now we fix a **Z**-basis $P := (\lambda_1, \cdots, \lambda_r)$ of **R**-independent λ_j $(j = 1, \cdots, r)$. Let us denote by $A \in \mathrm{M}(r, \mathbf{Z})$ the matrix of the integral entries $A(\lambda_j, \lambda_k)$ $(j, k = 1, \cdots, r)$ of the characteristic alternating A on this basis and by $\chi := (\chi_{\lambda_1}, \cdots, \chi_{\lambda_r})$ the values of the characteristic homomorphism on $P = (\lambda_1, \cdots, \lambda_r)$. Then $\langle\chi_{\lambda_j}, \lambda_k\rangle - \langle\chi_{\lambda_k}, \lambda_j\rangle = A(\lambda_j, \lambda_k)$ so that

(PRC) $\qquad {}^t\chi P - {}^tP\chi = A \quad \text{for a} \quad \chi \in \mathrm{M}(n, r; \mathbf{C}).$

(Period relations for the characteristic homomorphism)

The solutions of this equality are exactly $\chi + SP$ with a special solution χ and any symmetric $S \in \mathrm{M}(n, \mathbf{C})$. The characteristic homomorphisms defined by SP on a fixed basis P of Λ define the analytically trivial line bundles $X \times \mathbf{C}$ on X as we have seen after the proof of Corollary 3.1.2.

In standard coordinates $\Lambda = \mathbf{Z}^n \oplus \Gamma$ the characteristic homomorphism for a Γ-reduced theta factor is uniquely determined as we have seen in the chararacteristic decomposition theorem, namely $\chi = (0, \chi_\Gamma)$ with $\chi_\Gamma \in \mathrm{M}(n, q; \mathbf{Z})$.
In the Hermitian decomposition theorem 3.1.1 we proved $\langle \chi_\lambda, z \rangle = \frac{1}{2i}(H + S)(z, \lambda)$ with a Hermitian form H and a symmetric bilinear form S. As before we denote their entries on the basis of the $\mathbf{C}^n$ by the same symbols H, S, respectively. Then $\chi = \frac{1}{2i}[H\overline{P} + SP]$ so that by (PRC)

(PRH) $\quad \mathrm{Im}\,{}^tPH\bar{P} = A \quad$ with a Hermitian matrix H

(Period relations for the Hermitian H)

of course without the symmetric part. As before the part $\frac{1}{2i}SP$ defines the analytically trivial bundle $X \times \mathbf{C}$. The Hermitian H is not uniquely determined for $q < n$ as we have seen in the Hermitian decomposition theorem, even not in the Γ-reduced case.

We can write the characteristic alternating $A \in \mathrm{M}(n + q, \mathbf{Z})$ in block form

$$A = \begin{pmatrix} A_1 & -A_2 \\ {}^tA_2 & A_3 \end{pmatrix} \quad \text{with } A_1 \in \mathrm{M}(n; \mathbf{Z}),\ A_3 \in \mathrm{M}(q; \mathbf{Z}).$$

If we take a standard basis $P = (I_n, G)$ of $\Lambda = \mathbf{Z}^n \oplus \Gamma$, then we get readily from (PRH)

(PRS) $\quad {}^tGA_1G + {}^tGA_2 - {}^tA_2G + A_3 = 0.$

(Period relations in standard coordinates)

From here we can go back to (PRC). For that we remark that $A_1 = 0$ for a Γ-reduced theta factor. Then we can take $\chi_{I_n} = 0, \chi_G := A_2$ and with $\chi := (\chi_{I_n}, \chi_G)$ we get (PRC) by using (PRS) only.

So most of the following theorem is proved:

3.1.5 Theorem (Period relations)

Each of the period relations (PRC), (PRH) and (PRS) is equivalent with the existence of a line bundle L on $X = \mathbf{C}^n/\Lambda$ defined by a reduced theta factor and with the Chern class $c_1(L)$ given by the integral and alternating matrix A.

Proof

Subsequently, we have to construct only a semi–character ϱ on Λ for a given A. Then obviously by (PRC) or (PRH)

$$\vartheta_\lambda(z) := \varrho(\lambda)\mathrm{e}(\langle\chi_\lambda, z\rangle + \langle\chi_\lambda, \lambda\rangle) \quad or \quad := \varrho(\lambda)\mathrm{e}(\tfrac{1}{2i}[H(z,\lambda) + H(\lambda,\lambda)])$$

are theta factors defining such a line bundle.

To construct a semi–character it is enough to construct real c_λ $(\lambda \in \Lambda)$ with

$$c_{\lambda+\lambda'} \equiv c_\lambda + c_{\lambda'} + \frac{1}{2}A(\lambda, \lambda') \bmod \mathbf{Z} \qquad (\lambda, \lambda' \in \Lambda).$$

Then clearly $\varrho(\lambda) := \mathrm{e}(c_\lambda)$ $(\lambda \in \Lambda)$ becomes a semi–character of the desired type.

For the construction, let $\lambda_1, \cdots, \lambda_r$ be a $\mathbf{R}$-independent $\mathbf{Z}$-basis of Λ. Start with $c_0 = 0$ and *any* collection $c_{\lambda_j} \in \mathbf{R}$ $(j = 1, \cdots, r)$. Then define

$$\begin{aligned} c_{m_1\lambda_1+\cdots+m_r\lambda_r} & \\ := m_1c_{\lambda_1} + \cdots + m_rc_{\lambda_r} &+ \tfrac{m_1m_2}{2}A(\lambda_1,\lambda_2) + \cdots + \tfrac{m_1m_r}{2}A(\lambda_1,\lambda_r) \\ &\ddots \\ &+ \tfrac{m_{r-1}m_r}{2}A(\lambda_{r-1},\lambda_r) \end{aligned}$$

and use $\frac{1}{2}A(\lambda_j,\lambda_k) \equiv \frac{1}{2}A(\lambda_k,\lambda_j) \bmod \mathbf{Z}$ $(1 \le j,k \le r)$ to see the required congruence. *Q.E.D.*

Now we can calculate the Néron–Severi group of a toroidal group of type q using the characteristic homomorphism χ. The theta factor which defines a line bundle on a toroidal group $X = \mathbf{C}^n/\Lambda$ with $\Lambda = \mathbf{Z}^n \oplus \Gamma$ is uniquely determined by the Γ–reduced characteristic homomorphism $\chi : \Gamma \to \mathbf{Z}^n$. So we take a basis $P = (I_n, G)$ of Λ in standard coordinates with a basis G of Γ. Finally we define $\chi \in \mathrm{M}(n,q;\mathbf{Z})$ as the values of the Γ–reduced characteristic homomorphism on G. We get

$$\mathrm{NS}(X) \simeq \{\chi \in \mathrm{M}(n,q;\mathbf{Z}) : {}^t\chi G - {}^tG\chi \in \mathrm{M}(q,\mathbf{Z})\}$$

for toroidal groups of type q.

Example: For $q = 1$ the condition in the bracket is empty so that we can identify the Γ–reduced characteristic homomorphisms with all $\chi \in \mathrm{M}(n,1;\mathbf{Z})$:

$$\mathrm{NS}(X) \simeq \mathbf{Z}^n \quad \text{for all toroidal groups } X \text{ of type } 1,$$

especially for 1–dimensional torus groups $\mathrm{NS}(X) \simeq \mathbf{Z}$.

Non–compact toroidal groups are natural $\mathbf{C}^{*n-q}$–fibre bundles over a q–dimensional torus T. We write $G = \binom{\hat{T}}{\tilde{T}}$ in standard coordinates with $\operatorname{Im}\hat{T} \in \mathrm{GL}(q,\mathbf{R})$ and take $\chi = \binom{\chi_T}{0}$ with $\chi_T \in \mathrm{M}(q;\mathbf{Z})$ so that

$$\mathrm{NS}(T) \simeq \{\chi_T \in \mathrm{M}(q;\mathbf{Z}) : {}^t\chi_T\hat{T} - {}^t\hat{T}\chi_T \in \mathrm{M}(q,\mathbf{Z})\} \subset \mathrm{NS}(X)$$

becomes the subgroup of all lifted line bundles.

For further calculations of the Néron–Severi group of toroidal groups see the papers of SELDER [97, 98].

Ample Riemann forms

Quasi-Abelian varieties are toroidal groups with an ample Riemann form. In the compact case those are exactly Abelian varieties. The general concept of Riemann forms we bring later in Section 4.1 on p 101.

3.1.6 Definition

An **ample Riemann form** H for a discrete subgroup $\Lambda \subset \mathbf{C}^n$ of complex rank n is a Hermitian form H on $\mathbf{C}^n$ such that

1. $\mathrm{Im}H|_{\Lambda\times\Lambda}$ is $\mathbf{Z}$-valued,
2. H is positive definite on the maximal $\mathbf{C}$–linear subspace $\mathrm{M}\mathbf{C}_\Lambda$ of $\mathbf{R}_\Lambda$.

A **quasi–Abelian variety** is a toroidal group $X = \mathbf{C}^n/\Lambda$ with an ample Riemann form for Λ.

The definition of a quasi–Abelian variety does not depend on the representation of $X = \mathbf{C}^n/\Lambda$ with a lattice $\Lambda \subset \mathbf{C}^n$:
Isomorphic toroidal groups are represented by lattices $C\Lambda$ with $C \in \mathrm{GL}(n;\mathbf{C})$ [see Hurwitz relations on p 8]. The basis P of Λ transforms to $P_C = CP$ and the Hermitian matrix H to the Hermitian matrix $H_C = {}^tC^{-1}H\overline{C}^{-1}$ belonging to the transformed lattice.
We can speak about ample Riemann forms of toroidal groups.
It is well known from elementary geometry that the imaginary part of a positive definite Hermitian form is nonsingular. So, if the Hermitian form H is positive definite on the maximal complex subspace $\mathrm{M}\mathbf{C}_\Lambda$ of a toroidal group of type q, then $A = \mathrm{Im}\ H$ has the rank at least $2q$.

We can easily substitute any ample Riemann form by one that is positive definite on the *whole* $\mathbf{C}^n$ [33, 87]:

3.1.7 Lemma

Let $\Lambda \subset \mathbf{C}^n$ be a discrete subgroup of complex rank n and H an ample Riemann form for Λ. Then there exists a Hermitian form $\tilde{H}$ on $\mathbf{C}^n$, which is symmetric on $\mathbf{R}_\Lambda$, such that $H + \tilde{H}$ is positive definite on $\mathbf{C}^n$ - of course with the same imaginary part on $\mathbf{R}_\Lambda$ as H.

Proof

As in the proof of the Hermitian decomposition 3.1.1 let us take toroidal coordinates so that $\mathbf{R}_\Lambda = \mathrm{M}\mathbf{C}_\Lambda \oplus V$. Take a symmetric $\mathbf{R}$-bilinear form R_V on $V \times V$,

defined by a real diagonal matrix with positive entries. R_V can be extended to an $\mathbf{R}$-valued and symmetric bilinear form R on $\mathbf{R}_\Lambda \times \mathbf{R}_\Lambda$ by $R := R_V$ on $V \times V$ and $:= 0$ elsewhere. This R has a unique Hermitian extension $\tilde{H}$ to $\mathbf{C}^n$. The determinants of the principal minors of the first j rows of the matrix $H + \tilde{H}$ are positive for $j := q+1, j := q+2, \cdots$, if the entries $\tilde{h}_{q+1,q+1}, \tilde{h}_{q+2,q+2}, \cdots$ of $\tilde{H}$ successively are chosen sufficiently big.
At least $H + \tilde{H}$ must be positive definite. *Q.E.D.*

The first aim is to extend the lattice of the given toroidal group to a lattice of a torus group. For this we prove the

3.1.8 Lemma
Let $\Lambda \subset \mathbf{C}^n$ be a lattice of complex rank n and real rank $r < 2n$, H a non-degenerate Hermitian form on $\mathbf{C}^n$ so that $\mathrm{Im}H$ is $\mathbf{Z}$-valued on $\Lambda \times \Lambda$ and $F \subset \mathbf{C}^n$ a countable subset. Then there exists a $\lambda_0 \in \mathbf{C}^n \setminus \{\mathbf{R}_\Lambda \cup F\}$ so that $\mathrm{Im}H$ is $\mathbf{Z}$-valued on $\Lambda' := \Lambda \oplus \mathbf{Z}\lambda_0$.
Proof
Let $P = (\lambda_1, \cdots, \lambda_r)$ be a basis of Λ with $r < 2n$. We add $\lambda_{r+1}, \cdots, \lambda_{2n} \in \mathbf{C}^n$ to a basis $\hat{P} = (\lambda_1, \cdots, \lambda_{2n})$ of a torus lattice $\hat{\Lambda} \supset \Lambda$.
Consider the $\mathbf{R}$–linear map

$$\hat{L}(\hat{x}) := \hat{A}\hat{x} \ (\hat{x} \in \mathbf{R}^{2n}), \quad \text{where} \quad \hat{A} := \mathrm{Im}\,{}^t PH\overline{\hat{P}}.$$

Here we used the same symbol H for the coefficient matrix.
Obviously it is enough to show the
Assertion. there exists an $x_0 \in \mathbf{R}^{2n}$ so that

$$\hat{L}(x_0) = \hat{A}x_0 \in \mathbf{Z}^r \quad \text{and} \quad \lambda_0 := \hat{P}x_0 \in \mathbf{C}^n \setminus \{\mathbf{R}_\Lambda \cup F\}.$$

Indeed, once the assertion has been proved, then $\mathrm{Im}\,{}^t P_0 H \overline{P_0}$ becomes an integral matrix, where $P_0 := (\lambda_1, \cdots, \lambda_r, \lambda_0)$. because the imaginary part of the Hermitian form H is alternating, *Proof of the assertion.* Let $L(x) := Ax$ $(x \in \mathbf{R}^r)$ with $A := \mathrm{Im}\,{}^t PH\overline{P} \in \mathrm{M}(r, \mathbf{Z})$. We note that $\hat{L} : \mathbf{R}^{2n} \to \mathbf{R}^r$ is surjective because the imaginary part of the nonsingular Hermitian H is nonsingular.
Case a) rank $A = r$. In this case $L(x) = \tau_0$ has the unique solution for any $\tau_0 \in \mathbf{Z}^r$. Then we can take any $x_0 \in \mathbf{R}^{2n}$ such that $\hat{L}(x_0) = \tau_0$ with $\lambda_0 = \hat{P}x_0 \notin \mathbf{R}_\Lambda \cup F$. This is possible because dim ker $\hat{L} = 2n - r > 0$ and F is a countable set.
Case b) rank $A < r$. Let $\tau_0 \in \mathbf{Z}^r$ be any integer not in the image of L. Then we can take a solution x_0 of $\hat{L}(x_0) = \tau_0$ and $\lambda_0 = \hat{P}x_0 \notin \mathbf{R}_\Lambda \cup F$ by the same reason as above. *Q.E.D.*

In the next step we extend the lattice of a quasi–Abelian variety to lattices of Abelian varieties.

3.1.9 Proposition
Let $X = \mathbf{C}^n/\Lambda$ be a quasi–Abelian variety of rank $q < n$ with an ample Riemann form H for Λ, which is positive definite on $\mathbf{C}^n$. Then there are lattices of torus groups Λ_1, Λ_2 with $\Lambda_1 \cap \Lambda_2 = \Lambda$ so that H becomes an ample Riemann form as well for Λ_1 as for Λ_2.
Proof
.By Lemma 3.1.8 there exist lattices Λ_1, Λ_2 of rank $n+q+1$ such that $\Lambda_1 \cap \Lambda_2 = \Lambda$. For that we can take any $\Lambda_1 = \Lambda \oplus \mathbf{Z}\lambda_1$ with $\lambda_1 \notin \mathbf{R}_\Lambda$ and $\Lambda_2 = \Lambda \oplus \mathbf{Z}\lambda_2$ with $\lambda_2 \notin \Lambda_1$ and not in the $\mathbf{Q}$-span of λ_1.
By induction over the rank r of the lattices $\Lambda_{r1}, \Lambda_{r2}$ we take care that the new chosen period is not in the $\mathbf{R}$-span of its lattice and not in the $\mathbf{Q}$-span of the basis periods of the the other one. *Q.E.D.*

GHERARDELLI arranged in the years 1971-73 a seminar about "Varieté Quasi–Abeliene" with results of ANDREOTTI and himself and published in 1974 two theorems nearly without proofs. The first theorem of their results is the

3.1.10 Theorem (GHERARDELLI–ANDREOTTI [32, 33])
A toroidal group is a quasi–Abelian variety, iff it is the covering group of an Abelian variety.
Proof
$\succ$. Let $X = \mathbf{C}^n/\Lambda$ be a quasi–Abelian variety and $\check{\Lambda}$ be a lattice of an Abelian variety $T = \mathbf{C}^n/\check{\Lambda}$ with $\Lambda \subset \check{\Lambda}$ (Proposition 3.1.9) so that the imaginary part of the ample Riemann form H for Λ, which is positive definite on $\mathbf{C}^n$, remains integral on $\check{\Lambda} \times \check{\Lambda}$. Then the identity $\hat{\iota} : \mathbf{C}^n \hookrightarrow \mathbf{C}^n$ with $\hat{\iota}(\Lambda) \subset \check{\Lambda}$ induces a covering homomorphism $\iota : \mathbf{C}^n/\Lambda \to \mathbf{C}^n/\check{\Lambda}$ by Hurwitz relations (p 8).
$\prec$. By Hurwitz relations the lift $\hat{\iota}$ of the covering map ι is a linear map with $\hat{\iota}(\Lambda) \subset \check{\Lambda}$ and an ample Riemann form for $\check{\Lambda}$ is an ample Riemann form for Λ.
Q.E.D.

Quasi–Abelian varieties are situated in the covering hierarchy between the Abelian varieties and the space $\mathbf{C}^{*n}$ of holomorphic Fourier series.

A consequence of Proposition 3.1.9 is the theorem of HEFEZ in 1978 [43], which was conjectured by ANDREOTTI. Also POTHERING [87] obtained in 1977 this theorem.

3.1.11 Theorem (HEFEZ)
The meromorphic function field $\mathcal{M}(X)$ of a non–compact quasi–Abelian variety X has infinite transcendental degree over $\mathbf{C}$.
Proof
Let $X = \mathbf{C}^n/\Lambda$ be a non–compact quasi–Abelian variety and $\check{\Lambda} \supseteq \Lambda$ a lattice of an Abelian variety according Proposition 3.1.9 with a common ample Riemann form H. Moreover let $P = (\lambda_1, \cdots, \lambda_r)$ be a standard basis of Λ and

a standard basis of $\check{\Lambda}$. Transforming both simultaneously with $-\overline{H}$ we get the new period relations

$$\mathrm{Im}^t\overline{B}H^{-1}B = A \text{ with } B := -H\check{\overline{P}}.$$

In the classical theory of theta functions the existence of *Jacobian functions* $h : \mathbf{C}^n \to \mathbf{C}$ defined by $h(z + \lambda_j) = \mathbf{e}(\langle b_j, z\rangle + c_j)\ h(z)\ (j = 1, \cdots, 2n)$ with $B =: (b_1, \cdots, b_{2n})$ *any* constants $c_1, \cdots, c_{2n} \in \mathbf{C}$ is well known.
For the following construction we start with sequences
$1, d_0, d_1, d_2, \cdots$ of $\mathbf{Q}$–linearly independent complex numbers and
$h, h_0, h_1, h_2, \cdots$ of Jacobian functions defined by the equations

$$h(z + \lambda_j) \quad = \mathbf{e}(\langle b_j, z\rangle)h(z) \qquad (j = 1, \cdots, 2n)$$

and

$$h_k(z + \lambda_j) \quad = \mathbf{e}(\langle b_j, z\rangle)h_k(z) \qquad (j = 1, \cdots, 2n-1)$$
$$h_k(z + \lambda_{2n}) = \mathbf{e}(\langle b_{2n}, z\rangle + d_k)h_k(z).$$

Then define the Λ-periodic meromorphic functions $f_k := h_k/h\ (k \in \mathbf{N})$, which have the additional property

$$f_k(z + l\lambda_{2n}) = \mathbf{e}(ld_k)f_k(z) \qquad (k \in \mathbf{N}).$$

Assertion. The $f_k\ (k \in \mathbf{N})$ are algebraically independent.
Proof of the assertion
Let $P(x) = \sum_{|\mu|\le\deg} a_\mu x^\mu \in \mathbf{C}[x]$ be a polynomial with $P \ne 0$ and $P(f_0, f_1, \cdots, f_m) = 0$. Put $d := (d_0, d_1, \cdots, d_m)$. Then
$P\big(f_0(z + l\lambda_{2n}), \cdots, f_m(z + l\lambda_{2n})\big) = \sum_{|\mu|\le\deg} a_\mu \mathbf{e}(l\langle\mu, d\rangle) f_0^{\mu_0}(z) \cdots f_m^{\mu_m}(z) \equiv 0.$
We can enumerate the N summands of $P(f_0, f_1, \cdots, f_n)$ with nonvanishing coefficients by

$$\tau : (1, \cdots, N) \to \{\mu \in \mathbf{N}^{m+1} : a_\mu \ne 0\},$$

so that the t-th summand becomes

$$S_t(z) = a_{\tau_t} f_0^{(\tau_t)_0}(z) \cdots f_m^{(\tau_t)_m}(z) \qquad (t = 1, \cdots, N).$$

For every fixed $z \in \mathbf{C}^n$ we have a system of equations

$$\sum_{t=1}^{N} \mathbf{e}(l\langle\tau_t, d\rangle)S_t(z) \equiv 0 \qquad (l = 0, 1, \cdots, N-1)$$

with a constant coefficient matrix

$$V := \begin{pmatrix} 1 & 1 & \cdots & 1 \\ \mathbf{e}(\langle\tau_1, d\rangle) & \mathbf{e}(\langle\tau_2, d\rangle) & \cdots & \mathbf{e}(\langle\tau_N, d\rangle) \\ \mathbf{e}(2\langle\tau_1, d\rangle) & \mathbf{e}(2\langle\tau_2, d\rangle) & \cdots & \mathbf{e}(2\langle\tau_N, d\rangle) \\ \vdots & \vdots & & \vdots \\ \mathbf{e}((N-1)\langle\tau_1, d\rangle) & \mathbf{e}((N-1)\langle\tau_2, d\rangle) & \cdots & \mathbf{e}((N-1)\langle\tau_N, d\rangle) \end{pmatrix}.$$

But all the S_t cannot vanish simultaneously everywhere in the $\mathbf{C}^n$ so that Vandermonde's determinant has to vanish

$$\det V = (-1)^{\frac{N(N-1)}{2}} \prod_{1 \le j < k \le n} \Big(\mathbf{e}(\langle \tau_j, d \rangle) - \mathbf{e}(\langle \tau_k, d \rangle) \Big) = 0.$$

At least one factor has to vanish so that $\langle \tau_{j_0}, d \rangle \equiv \langle \tau_{k_0}, d \rangle \bmod \mathbf{Z}$ for certain $j_0 < k_0$. The $1, d_0, d_1, \cdots, d_m$ become $\mathbf{Q}$–linearly dependent. *Q.E.D.*

Maximal Stein subgroups of quasi–Abelian varieties

In the last section we have seen that the imaginary part of an ample Riemann form has at least rank $2q$. The following definition is due to GHERARDELLI and ANDREOTTI [33].

3.1.12 Definition
Let $X = \mathbf{C}^n/\Lambda$ be a quasi–Abelian variety of type q.
An ample Riemann form H for Λ is said to be of **kind ℓ**, iff $\mathrm{Im}H$ has the rank $2q + 2\ell$ on $\mathbf{R}_\Lambda$.

For a fixed lattice $\Lambda \subset \mathbf{C}^n$ we can take a special basis so that $A = \mathrm{Im}H$ becomes the wellknown Frobenius normal form:

3.1.13 Lemma (FROBENIUS)
For any alternating $A \in \mathrm{M}(r, \mathbf{Z})$ of rank $2k$ there exist a unimodular matrix $M \in \mathrm{GL}(r, \mathbf{Z})$ and positive integers $d_1 | d_2 | \cdots | d_k$ so that A becomes *unimodular–equivalent* to the **Frobenius normal form**

$${}^tMAM = \begin{pmatrix} 0 & D & 0 \\ -D & 0 & 0 \\ 0 & 0 & 0 \end{pmatrix} \quad \text{with} \quad D = \mathrm{diag}(d_1, \cdots, d_k).$$

The integers $d_1, \cdots, d_k$ are the **elementary divisors** of A.

For the *proof* of this classical lemma see GRÖBNER [40] or LANG [65].

By Hurwitz relations (p 8) we can assume that the matrix $A := \mathrm{Im}\,{}^tPH\overline{P}$ of the characteristic alternating form $A = \mathrm{Im}H$ is given in its Frobenius normal form on a special basis P of Λ.

A $\mathbf{C}$–linear map $\hat{\tau} : \mathbf{C}^n \to \mathbf{C}^{n'}$ induces a contravariant map $\tau^* : \mathrm{Her}(\mathbf{C}^{n'}) \to \mathrm{Her}(\mathbf{C}^n)$ of the Hermitian forms by $\tau^*(H')(x, y) := H'(\tau(x), \tau(y))$ $(x, y \in \mathbf{C}^n)$.

3.1.14 Definition
Let X be a toroidal group and X' a quasi–Abelian variety.
A surjective complex homomorphism $\tau : X \to X'$ is a **homomorphism pre-**

serving an ample Riemann form H' of X', if its preimage $\tau^*(H')$ is an ample Riemann form of X.

The **kernel** of a bilinear or Hermitian form on a vector space V is defined as

$$\operatorname{Ker} H := \{v \in V : H(v,w) = 0 \text{ for all } w \in V\}.$$

3.1.15 Proposition
For a surjective complex homomorphism τ of a toroidal group X onto a quasi–Abelian variety X' the following statements are equivalent:

1. τ **is a homomorphism preserving *one* ample Riemann form.**
2. τ **is a homomorphism preserving *all* ample Riemann forms, which are positive on $\mathbf{C}^n$.**
3. **The kernel of τ is a connected Stein subgroup.**

Proof
Let $X = \mathbf{C}^n/\Lambda$ be a toroidal group and $X' = \mathbf{C}^{n'}/\Lambda'$ a quasi–Abelian variety.
1≻3. Let the preimage $H := \tau^*(H')$ of an ample Riemann form H' for Λ' be an ample Riemann form for Λ.
Moreover let $E \subset \mathbf{C}^n$ be the universal covering space of $\ker\tau \subset X$. Then $E \subset \operatorname{Ker} H$ so that $H = 0$ on $(E \cap \mathbf{R}_\Lambda) \times (E \cap \mathbf{R}_\Lambda)$. Let F be the maximal $\mathbf{C}$–linear subspace of $E \cap \mathbf{R}_\Lambda$. Then $F \subset \mathbf{MC}_\Lambda$. But $H > 0$ on $\mathbf{MC}_\Lambda$ so that $F = 0$. By Stein group criterion for Abelian Lie groups (Lemma 1.1.6) $\ker\tau \simeq E/(E \cap \Lambda)$ must be a connected Stein group.
3≻2. Let H' be any ample Riemann form for Λ', which is positive definite on $\mathbf{C}^n$, H its preimage and the kernel $\ker\tau$ a Stein subgroup of X with universal covering group E. Then by Stein group criterion 1.1.6 $E \cap \mathbf{R}_\Lambda$ has no complex subspace of positive dimension so that $E \cap \mathbf{MC}_\Lambda = 0$. E is the kernel of the lift $\hat{\tau}$, therefore $=\operatorname{Ker} H$ so that H is positive definite on $\mathbf{MC}_\Lambda$. *Q.E.D.*

Of course the preimage of a homomorphism preserving an ample Riemann form is a quasi–Abelian variety. However, the image of a quasi–Abelian variety under a homomorphism need not be quasi–Abelian, even if the kernel is a connected Stein group according the previous proposition, as an example of Abe [6] shows.

Example: The basis

$$P = \begin{pmatrix} 1\,0\,0 & 0 & 5i \\ 0\,1\,0 & i & 0 \\ 0\,0\,1 & \sqrt{2}+\sqrt{3}i & \sqrt{7} \end{pmatrix}$$

generates a toroidal group X, which allows two projections to the 2–dimensional torus groups generated by

$$P_1 = \begin{pmatrix} 1\,0\,0\,5i \\ 0\,1\,i\,\,0 \end{pmatrix} \quad \text{and} \quad P_2 = \begin{pmatrix} 1\,0 & 0 & 5i \\ 0\,1 & \sqrt{2}+\sqrt{3}i & \sqrt{7} \end{pmatrix}, \quad \text{respectively.}$$

The torus generated by P_1 is proective algebraic, because it is the product of two groups of dimension one. The kernel of the first projection is isomorphic to $\mathbf{C}^*$. Therefore X is quasi-Abelian. But the torus generated by P_2 is not algebraic because the period relations are not fulfilled, - even if the kernel of the second projection is also isomorphic to a $\mathbf{C}^*$.

One main result of the seminar of GHERARDELLI and ANDREOTTI was Theorem 2 of [33]. They called it *Fibration Theorem.* DODSON brought a proof in 1980 together with other results [27]. We follow ABE [6].

3.1.16 Fibration Theorem (GHERARDELLI–ANDREOTTI)
Let $X = \mathbf{C}^n/\Lambda$ be a toroidal group of type q. Then are equivalent:

1. X is a quasi-Abelian variety with an ample Riemann form for Λ of kind ℓ.
2. X has a maximal closed Stein subgroup $N \simeq \mathbf{C}^\ell \times \mathbf{C}^{*m}$ with $2\ell + m = n - q$ and X/N. is an Abelian variety of dimension $q + \ell$.

Proof
$2 \succ 1.$ Proposition 3.1.15 because the projection is a homorphism preserving an ample Riemann form.
$1 \succ 2.$ Let H be positive definite on $\mathbf{C}^n$ (Lemma 3.1.7).
1st step. Using Frobenius' Lemma we can choose a basis $P = (\lambda_1, \cdots, \lambda_{n+q})$ of Λ such that with $m := n - q - 2\ell$

$$\operatorname{Im}{}^t PH\overline{P} = \begin{pmatrix} 0 & 0 & D \\ 0 & 0 & 0 \\ -D & 0 & 0 \end{pmatrix} \}m \text{ rows} \quad \text{with} \quad D = \operatorname{diag}(d_1, \cdots, d_{q+\ell}).$$

If $m > 0$ we can select in the same way as in the proof of Lemma 3.1.8 a further period λ_{n+q+1} so that with the new basis $P_1 = (\lambda_1, \cdots, \lambda_{n+q+1})$ of the extended lattice Λ_1

$$\operatorname{Im}{}^t P_1 H\overline{P}_1 = \begin{pmatrix} 0 & 0 & D_1 \\ 0 & 0 & 0 \\ -D_1 & 0 & 0 \end{pmatrix} \}m\text{-1 rows}, \quad \text{where } D_1 = \operatorname{diag}(d_1, \cdots, d_{q+\ell+1})$$

remains nonsingular and integral. Continuing this procedure till $m = 0$, we get a basis $P_m = (\lambda_1, \cdots, \lambda_{2n-2\ell})$ of a lattice Λ_m so that integral valued

$$\operatorname{Im}{}^t P_m H\overline{P}_m = \begin{pmatrix} 0 & D_m \\ -D_m & 0 \end{pmatrix}, \quad \text{with } D_m = \operatorname{diag}(d_1, \cdots, d_{n-l}),$$

becomes nonsingular.
2nd step. If $\ell > 0$, we get in the same way as in the proof of Lemma 3.1.8 a new

period μ_1 so that with the basis $P_{m+1} = (\lambda_1, \cdots, \lambda_{n-\ell}, \mu_1, \lambda_{n-\ell+1}, \cdots, \lambda_{2n-2\ell})$ of the extended lattice Λ_{m+1}

$$\operatorname{Im} {}^tP_{m+1}H\overline{P}_{m+1} = \begin{pmatrix} 0 & E_1 \\ -{}^tE_1 & 0 \end{pmatrix}, \text{ with } E_1 = \begin{pmatrix} d_1 & \cdots & 0 \\ \vdots & \ddots & \vdots \\ 0 & \cdots & d_{n-\ell} \\ 0 & \cdots & 0 \end{pmatrix},$$

is integral valued. Then we get by the same lemma a new period γ_1 so that with the basis $P_{m+2} = (\lambda_1, \cdots, \lambda_{n-\ell}, \mu_1, \lambda_{n-\ell+1}, \cdots, \lambda_{2n-2\ell}, \gamma_1)$ of the extended lattice Λ_{m+2}

$$\operatorname{Im} {}^tP_{m+2}H\overline{P}_{m+2} = \begin{pmatrix} 0 & E_2 \\ -{}^tE_2 & 0 \end{pmatrix}, \text{ with } E_2 = \begin{pmatrix} d_1 & \cdots & 0 & 0 \\ \vdots & \ddots & \vdots & \vdots \\ 0 & \cdots & d_{n-\ell} & 0 \\ 0 & \cdots & 0 & d_{n-\ell+1} \end{pmatrix},$$

is nonsingular and integral. Finally there exists a basis of a torus group $\check{P} := P_{m+2\ell} = (\lambda_1, \cdots, \lambda_{n-\ell}, \mu_1, \cdots, \mu_\ell, \lambda_{n-\ell+1}, \cdots, \lambda_{2n-2\ell}, \gamma_1, \cdots, \gamma_\ell)$ of the extended lattice $\check{\Lambda} = \Lambda_{m+2\ell}$. With integral and nonsingular $\check{E} := E_{m+2\ell}$ we get

$$\operatorname{Im} {}^t\check{P}H\overline{\check{P}} = \begin{pmatrix} 0 & \check{E} \\ -{}^t\check{E} & 0 \end{pmatrix} \text{ where } \check{E} = \begin{pmatrix} d_1 & & & & 0 \\ & \ddots & & & \\ & & d_{n-l} & & \\ & & & \ddots & \\ 0 & & & & d_n \end{pmatrix}.$$

$\check{X} := \mathbf{C}^n / \check{\Lambda}$ is an Abelian variety with the same Riemann form as X.
3rd step. As a consequence $\check{P}$ is a basis of the Abelian variety $\check{X} = C^n/\check{\Lambda}$, H a Riemann form for $\check{\Lambda}$ and

$$\check{A} := \operatorname{Im} {}^t\check{P}H\overline{\check{P}} = \begin{pmatrix} 0 & \check{E} \\ -{}^t\check{E} & 0 \end{pmatrix} \text{ integral and nonsingular.}$$

In this situation we define $\check{P} =: (U, V)$ with quadratic U, V. Then there are two consequences of the classical period relations $\check{P}\check{A}^{-1}\,{}^t\check{P} = 0$ and $i\overline{\check{P}}\check{A}^{-1}\,{}^t\check{P} > 0$:
1. U is regular and with $C :={}^t \check{E}U^{-1}$
2. $C\check{P} = ({}^t\check{E}, W)$ with symmetric W and $\operatorname{Im} W > 0$.
$C\check{P}$ is a basis of $C\check{\Lambda}$ which defines the same Abelian variety as $\check{\Lambda}$. CP is the corresponding part in $C\check{P}$. We get

$$CP = \begin{pmatrix} d_1 & & & & & 0 & w_{1,1} & \cdots & w_{1,q+\ell} \\ & \ddots & & & & & \vdots & & \vdots \\ & & d_{q+l} & & & & w_{q+l,1} & \cdots & w_{q+\ell,q+\ell} \\ & & & \ddots & & & \vdots & & \vdots \\ & & & & d_{n-\ell} & & w_{n-\ell,1} & \cdots & w_{n-\ell,q+\ell} \\ & & & & & & \vdots & & \vdots \\ 0 & & & & & & w_{n,1} & \cdots & w_{n,q+\ell} \end{pmatrix}.$$

Because W is symmetric and $\mathrm{Im}W$ positive definite the principal minor

$$W^* = \begin{pmatrix} w_{1,1} & \cdots & w_{1,q+\ell} \\ \vdots & & \vdots \\ w_{q+\ell,1} & \cdots & w_{q+\ell,q+\ell} \end{pmatrix}$$

has the same properties.

If we define $T := (D, W^*)$ with $D = \mathrm{diag}(d_1, \cdots, d_{q+\ell})$ as above, then T generates the lattice of an $q+\ell$–dimensional Abelian variety. The projection onto the first $q+l$ variables induces a homomorphism $X \to Y := \mathbf{C}^n / T\mathbf{Z}^{2q+2\ell}$ according matrix CP with kernel $N \simeq \mathbf{C}^\ell \oplus \mathbf{C}^{*m}$. *Q.E.D.*

The theorem shows the importance of fibre bundles with Stein fibres and a projective algebraic base space.

3.1.17 Definition

A complex manifold X is **quasi–projective**, iff it is a submanifold of a projective algebraic variety $\overline{X}$ such that $\overline{X} \setminus X$ is an analytic set in $\overline{X}$.

A consequence of the Fibration Theorem is the

3.1.18 Theorem (Abe [4, 5])

Every quasi–Abelian variety is quasi–projective.

Proof

The fibration theorem shows that the quasi–Abelian variety $X = \mathbf{C}^n/\Lambda$ is a fibre bundle over an Abelian variety with Stein fibres $\simeq \mathbf{C}^\ell \otimes \mathbf{C}^{*m}$. We can compactify these fibres to $\mathbf{P}_{\ell+m}$ so that X becomes a submanifold in a fibre bundle $\overline{X}$ with an Abelian variety as base space and fibres $\mathbf{P}_{\ell+m}$.

A consequence of Kodaira's embedding theorem is that every fibre bundle with a projective algebraic base space, projective space fibres and projective general linear group as structure group is projective algebraic [60, Theorem 8 of Kodaira]. *Q.E.D.*

3.2 Characterization of quasi–Abelian varieties

The Hodge decomposition of de Rham cohomology groups works only for theta groups. In the general case the average of the Chern forms of a positive line bundle corresponds to the ample Riemann forms determined by the same line bundle. Holomorphic mappings to the complex projective spaces allow a straight forward proof of the Main Theorem which characterizes the quasi–Abelian varieties.

Cohomology and the average of differential forms

Let $X = \mathbf{C}^n/\Lambda$ be an Abelian Lie group of type q. The basis of Λ can be given in toroidal coordinates as

$$P = \begin{pmatrix} 0 & I_q & \hat{T} \\ I_{n-q} & R_1 & R_2 \end{pmatrix}$$

where $B =: (I_q, \hat{T})$ is the basis of a torus T and $R := (R_1, R_2)$ the glueing matrix. So the parametrization $L_{\mathbf{R}}(t) = At$ ($t \in \mathbf{R}^{2n}$) of the real written toroidal coordinates with

$$(L_{\mathbf{R}}) \qquad A := \begin{pmatrix} I_q & \mathrm{Re}\hat{T} & 0 & 0 \\ 0 & \mathrm{Im}\hat{T} & 0 & 0 \\ R_1 & R_2 & I_{n-q} & 0 \\ 0 & 0 & 0 & I_{n-q} \end{pmatrix}$$

transforms Λ-periodic functions into functions which are $\mathbf{Z}^{n+q}$-periodic in the first $n+q$ variables t' of $t = (t', t'') \in \mathbf{R}^{2n}$.

A Λ–periodic r-form can be written in the real t–parameter as

$$\omega(t) = \frac{1}{r!} \sum_{1 \le i_1, \cdots, i_r \le 2n} f_{i_1 \cdots i_r}(t) dt_{i_1} \wedge \cdots \wedge dt_{i_r} \qquad (t \in \mathbf{R}^{2n})$$

with $\mathbf{Z}^{n+q}$-periodic coefficients, which have Fourier series expansions

$$f_{i_1 \cdots i_r}(t) = \sum_{\sigma \in \mathbf{Z}^{n+q}} f^{(\sigma)}_{i_1 \cdots i_r}(t'') \, \mathbf{e}(\langle \sigma, t' \rangle).$$

3.2.1 Lemma (Abe)
The Fourier series $\sum_{\sigma \in \mathbf{Z}^{n+q}} f^{(\sigma)}(t'') \, \mathbf{e}(\langle \sigma, t' \rangle)$ converges to a $\mathbf{Z}^{n+q}$–periodic C^∞–function on $\mathbf{R}^{2n}$, iff for all $s \in \mathbf{R}$ and all multiindices I

$$\sum_{\sigma \in \mathbf{Z}^{n+q}} (1 + |\sigma|^2)^s \left| \frac{\partial^{|I|} f^{(\sigma)}}{\partial t''^I} \right|^2$$

converges uniformly on compact subsets of $\mathbf{R}^{n-q}$.

Proof

$\succ$. Let f be a $\mathbf{Z}^{n+q}$-periodic $\mathcal{C}^\infty$-function and

$$f(t) = \sum_{\sigma \in \mathbf{Z}^{n+q}} f^{(\sigma)}(t'')\,\mathbf{e}(\langle\sigma, t'\rangle).$$

Then for any $\mu \in \mathbf{Z}^{n+q}$, we have

$$\int_{\mathbf{T}} f(t)\mathbf{e}(-\langle\mu, t'\rangle)dt' = f^{(\mu)}(t''),$$

where $\mathbf{T} := (\mathbf{R}/\mathbf{Z})^{n+q}$. The same equality holds for all $D^I f := \partial^{|I|}/\partial t''^I f$ with any multiindex $I \geq 0$. By partial integration

$$\begin{aligned}(D_j f)^{(\mu)}(t'') &= \int_{\mathbf{T}} (D_j f)(t)\mathbf{e}(-\langle\mu, t'\rangle)dt' \\ &= -\int_{\mathbf{T}} D_j \mathbf{e}(-\langle\mu, t'\rangle)dt' = 2\pi i \mu_j f^{(\mu)}(t'')\end{aligned}$$

and in the same way

$$(D_j^2 f)^{(\mu)}(t'') = -4\pi^2\mu_j^2 f^{(\mu)}(t'').$$

By Parseval's equality applied on $F := (4\pi^2 - \sum D_j^2)^k f$ for any integer $k \geq 0$ we get

$$(4\pi^2)^2 \sum_{\sigma} (1 + |\mu|^2)^k \left|f^{(\mu)}(t'')\right|^2 = \int_{\mathbf{T}} \left|(4\pi^2 - D_1^2 - \cdots - D_{n+q}^2)^k f(t)\right|^2 dt'.$$

The right-hand side is continuous in t''. So by Dini's Theorem the series on the left-hand side converges uniformly on every compact subset $C \subset \mathbf{R}^{n-q}$. Of course this is true even for any $D^I f$.

$\prec$. By Sobolev's embedding theorem [72, p 78/79] the series

$$\sum_{\sigma} D^I f^{(\sigma)}(t'')\mathbf{e}(\langle\sigma, t'\rangle)$$

converges absolutely and uniformly on $\mathbf{T} \times C$, where C is any compact subset of $\mathbf{R}^{n-q}$ so that f is a $\mathcal{C}^\infty$-function. *Q.E.D.*

Now let $K := \mathbf{R}_\Lambda/\Lambda$ be the *maximal real compact subgroup* of X and T_K the *tangent bundle* of K. The tangent bundle T_X of X is identified with $T_K \times T_{\mathbf{R}^{n-q}}$ in $\mathcal{C}^\infty$-category, where $T_{\mathbf{R}^{n-q}}$ is the tangent bundle of $\mathbf{R}^{n-q}$.

3.2.2 Definition
Let $X = \mathbf{C}^n/\Lambda$ be an Abelian Lie group with Λ of rank $n+q$ and the maximal real subtorus $K = \mathbf{R}_\Lambda/\Lambda$. Then for any $\mathcal{C}^\infty$ r–form ω the **average of** ω **on** K is in toroidal coordinates defined by

$$Av(\omega)(t'') := \frac{1}{r!} \sum_{1 \le i_1 \cdots, i_r \le 2n} \Big(\int_K f_{i_1 \cdots i_r}(t', t'') dt' \Big) dt_{i_1} \wedge \cdots \wedge dt_{i_r} \quad (t'' \in \mathbf{R}^{n-q}).$$

Remark
1. The average of a $\mathcal{C}^\infty$–form is a $\mathbf{C}^\infty$–form with coefficients which depends in toroidal coordinates on the last $n-q$ variables only.
2. For any $\mathcal{C}^\infty$ (m,p)-form

$$\omega(t) = \frac{1}{m!}\frac{1}{p!} \sum_{\substack{|I|=m \\ |J|=p}} f_{IJ} dz_I \wedge d\bar{z}_J$$

we can define similarly the average $Av(\omega)(t'')$ of ω on K.

For a $\mathcal{C}^\infty$ r–form ω we define the r–form

$$\omega^{(\sigma)}(t) := \frac{1}{r!} \sum_{1 \le i_1, \cdots, i_r \le 2n} f^{(\sigma)}_{i_1 \cdots i_r}(t'')\, \mathbf{e}(\langle \sigma, t' \rangle) dt_{i_1} \wedge \cdots \wedge dt_{i_r}$$

so that

$$\omega = \sum_{\sigma \in \mathbf{Z}^{n+q}} \omega^{(\sigma)}.$$

In this case we obtain obviously

$$(Av) \qquad Av(\omega)(t'') = \omega^{(0)}(t) = \frac{1}{r!} \sum_{1 \le i_1, \cdots, i_r \le 2n} f^{(0)}_{i_1 \cdots i_r}(t'')\, dt_{i_1} \wedge \cdots \wedge dt_{i_r}.$$

Let $\mathcal{E}, \mathcal{E}^r$ be the sheaf of germs of complex valued $\mathcal{C}^\infty$–functions or $\mathcal{C}^\infty$ r–forms on X. By Poincaré lemma there exists an exact sequence

$$0 \to \mathbf{C} \xrightarrow{\iota} \mathcal{E}^0 \xrightarrow{d} \mathcal{E}^1 \xrightarrow{d} \mathcal{E}^2 \xrightarrow{d} \cdots \xrightarrow{d} \mathcal{E}^{2n} \to 0,$$

which induces the de Rham sequence of global sections so that

$$H^r(X, \mathbf{C}) \simeq Z(X, \mathcal{E}^r)/B(X, \mathcal{E}^r)$$

with the space $Z(X, \mathcal{E}^r) = \ker(\mathcal{E}^r(X) \xrightarrow{d} \mathcal{E}^{r+1}(X))$ of d-closed and the space $B(X, \mathcal{E}^r) = \mathrm{im}(\mathcal{E}^{r-1}(X) \xrightarrow{d} \mathcal{E}^r(X))$ of d–exact $\mathcal{C}^\infty$ r–forms on X.

Two r-forms ω_1, ω_2 are *d-cohomologous* ($\omega_1 \sim \omega_2$) iff their difference is d–exact.

3.2.3 Lemma
Let $X = \mathbf{C}^n/\Lambda$ be an Abelian Lie group with Λ of rank $n+q$. Then for any $\mathcal{C}^\infty$ r–form ω on X we have:

1. The average of ω is $Av(\omega) = \omega^{(0)}$.
2. If ω is d–exact, then $Av(\omega)(t'') = 0$ on $(T_K)^r$, i.e.

$$Av(\omega)(t'') = \frac{1}{r!} \sum_{\substack{1 \le i_1 \cdots, i_r \le 2n \\ n+q+1 \le i_l \le 2n \text{ for some } l}} f^{(0)}_{i_1 \cdots i_r}(t'') dt_{i_1} \wedge \cdots \wedge dt_{i_r}.$$

3. If ω is d–closed and $Av(\omega) = 0$, then ω is d–exact.
4. If ω is d-closed, then ω is d-cohomologous to $Av(\omega)$.

Proof
1. As seen before.
2. Let ω be d–exact such that $\omega = d\eta$ with a $\mathcal{C}^\infty$ $(r$-$1)$–form η and

$$\eta^{(\sigma)}(t) = \frac{1}{(r-1)!} \sum_{1 \le i_1, \cdots, i_{r-1} \le 2n} g^{(\sigma)}_{i_1 \cdots i_{r-1}}(t'')\, \mathbf{e}(\langle \sigma, t' \rangle) dt_{i_1} \wedge \cdots \wedge dt_{i_{r-1}}.$$

Then

(1) $f^{(\sigma)}_{i_1, \cdots, i_p}(t'')$

$$= \textstyle\sum_{k=1}^{l} (-1)^{k+1} 2\pi i\, \sigma_k\, g^{(\sigma)}_{i_1 \cdots \hat{i_k} \cdots i_r}(t'') + \sum_{k=l+1}^{r} (-1)^{k+1} \frac{\partial g^{(\sigma)}_{i_1 \cdots \hat{i_k} \cdots i_r}}{\partial t_{i_k}}(t''),$$

where $l := \max\{k : i_k \le n+q\}$.
Especially, if $1 \le i_1 < \cdots < i_r \le n+q$ and $\sigma = 0$, then

$$f^{(0)}_{i_1 \cdots i_r}(t'') \equiv 0. \tag{2}$$

3. Now let ω be d-closed so that $d\omega^{(\sigma)} = 0$ $(\sigma \in \mathbf{Z}^{n+q})$.
For every $\sigma \in \mathbf{Z}^{n+q} \setminus \{0\}$ we set $i(\sigma) := \max\{i : \sigma_i \ne 0\}$ and $S(\sigma) := \sigma_{i(\sigma)}$. It follows from (1) that for any multiindex $(i_1, \cdots, i_{r+1})$ with $1 \le i_1, \cdots, i_{r+1} \le 2n$

$$0 = \sum_{k=1}^{l} (-1)^{k+1} 2\pi i\, \sigma_k\, f^{(\sigma)}_{i_1 \cdots \hat{i_k} \cdots i_{r+1}}(t'') + \sum_{k=l+1}^{r+1} (-1)^{k+1} \frac{\partial f^{(\sigma)}_{i_1 \cdots \hat{i_k} \cdots i_{r+1}}}{\partial t_{i_k}}(t''),$$

where $l := \max\{k : i_k \le n+q\}$ as above.
For a fixed multiindex $(i_1, \cdots, i_r)$ with $1 \le i_1, \cdots, i_r \le 2n$ we consider the multiindex $(i(\sigma), i_1, \cdots, i_r)$. We get

$$2\pi i\, S(\sigma)\, f^{(\sigma)}_{i_1 \cdots i_r}(t'')\, \mathbf{e}(\langle \sigma, t' \rangle) = \sum_{k=1}^{r} (-1)^{k+1} \frac{\partial [f^{(\sigma)}_{i(\sigma) i_1 \cdots \hat{i_k} \cdots i_r}(t'')\, \mathbf{e}(\langle \sigma, t' \rangle)]}{\partial t_{i_k}}. \tag{3}$$

With

$$g^{(\sigma)}_{i_1 \cdots i_{r-1}}(t'') := \frac{f^{(\sigma)}_{i(\sigma) i_1 \cdots i_{r-1}}(t'')}{2\pi i\, S(\sigma)}$$

we define

$$(4)\qquad \eta^{(\sigma)} := \frac{1}{(r-1)!} \sum_{1\leq i_1,\cdots,i_{r-1}\leq 2n} g^{(\sigma)}_{i_1\cdots i_{r-1}}(t'')\,\mathrm{e}(\langle\sigma,t'\rangle)dt_{i_1}\wedge\cdots\wedge dt_{i_{r-1}}.$$

By (3) we get

$$f^{(\sigma)}_{i_1\cdots i_r}(t'')\,\mathrm{e}(\langle\sigma,t'\rangle) = \sum_{k=1}^{r}(-1)^{k+1}\frac{\partial[g^{(\sigma)}_{i_1\cdots\hat{i}_k\cdots i_r}(t'')\,\mathrm{e}(\langle\sigma,t'\rangle)]}{\partial t_{i_k}}$$

so that by (1)

$$\omega^{(\sigma)} = d\eta^{(\sigma)}.$$

By Lemma 6.1 and (4) $\eta := \sum_{\sigma\in\mathbf{Z}^{n+q}\setminus\{0\}}\eta^{(\sigma)}$ converges to an element of $H^0(X,\mathcal{E}^{r-1})$ and satisfies $d\eta = \sum_{\sigma\in\mathbf{Z}^{n+q}\setminus\{0\}}\omega^{(\sigma)}$. Since $\omega^{(0)} = Av(\omega) = 0$, we have $\omega = d\eta$.

4. $Av(\omega - Av(\omega)) = Av(\omega) - Av(\omega) = 0$. So by 3) $\omega - Av(\omega)$ is d–exact. *Q.E.D.*

As we have seen in Proposition 2.2.5 the following proposition is not true in general for $\overline{\partial}$–closed $(0,p)$–forms.

3.2.4 Proposition

Let $X = \mathbf{C}^n/\Lambda$ with Λ of rank $n+q$ and ω be a C^∞ r–form on X.
If ω is d-closed, then there exists a uniquely determined r–form χ on X with constant coefficients

$$\chi = \frac{1}{r!}\sum_{1\leq i_1,\cdots,i_r\leq n+q} c_{i_1\cdots i_r}dt_{i_1}\wedge\cdots\wedge dt_{i_r},$$

which is d-cohomologous to ω. χ is the sum of terms of $Av(\omega)$ containing only $dt_1,\cdots,dt_{n+q}$.

Proof

Let ω be d-closed.

Existence. In the sum $\omega = \sum\omega^{(\sigma)}$ all $\omega^{(\sigma)}$ are closed. We set $d =: d_{t'} + d_{t''}$. Obviously $d_{t'}\omega^{(0)} = 0$. Then $d_{t''}\omega^{(0)} = 0$ for $d\omega^{(0)} = 0$.

We decompose

$$\begin{aligned}\omega^{(0)} &= \frac{1}{r!}\sum_{1\leq i_1,\cdots,i_r\leq n+q} f^{(0)}_{i_1\cdots i_r}(t'')dt_{i_1}\wedge\cdots\wedge dt_{i_r}\\ &+\frac{1}{(r-1)!}\sum_{1\leq i_1,\cdots,i_{r-1}\leq n+q}\omega^{(0)}_{i_1\cdots i_{r-1}}\wedge dt_{i_1}\wedge\cdots\wedge dt_{i_{r-1}}\\ &\cdots\\ &+\sum_{1\leq i\leq n+q}\omega^{(0)}_i\wedge dt_i + \omega^{(0)}_0,\end{aligned}$$

where $\omega^{(0)}_{i_1\cdots i_{r-1}}, \cdots, \omega^{(0)}_i, \omega^{(0)}_0$ are a 1–form, $\cdots$, (r-1)–form, r-form, respectively, with respect to the differentials $dt_{n+q+1}, \cdots, dt_{2n}$. Considering the degree with respect to the same differentials, we obtain by $d_{t''}\omega^{(0)} = 0$ so that $d_{t''} f^{(0)}_{i_1\cdots i_r} = 0$, $d_{t''}\omega^{(0)}_{i_1\cdots i_{r-1}} = 0, \cdots, d_{t''}\omega^{(0)}_i = 0$ and $d_{t''}\omega^{(0)}_0 = 0$. Therefore $f^{(0)}_{i_1\cdots i_r}$ is constant $=: c_{i_1\cdots i_r}$ and by Poincaré's Lemma the $\omega^{(0)}_{i_1\cdots i_{r-1}}, \cdots, \omega^{(0)}_i$ and $\omega^{(0)}_0$ are $d_{t''}$-exact. Hence we set

$$\chi := \frac{1}{r!} \sum_{1\le i_1,\cdots,i_r \le n+q} c_{i_1\cdots i_r}\, dt_{i_1} \wedge \cdots \wedge dt_{i_r}$$

and we have $\omega^{(0)} = \chi + d\eta^{(0)}$ with a $\mathcal{C}^\infty$ (r-1)–form $\eta^{(0)}$.
Uniqueness. If $\omega^{(0)} = \chi_1 + d\eta^{(0)}_1 = \chi_2 + d\eta^{(0)}_2$, then $\chi_1 - \chi_2 = d(\eta^{(0)}_2 - \eta^{(0)}_1)$ so that by (2) $\chi_1 - \chi_2 = 0$. *Q.E.D.*

The cohomology class of any differential form in $Z(X, \mathcal{E}^r)$ is represented by a uniquely determined

$$\chi = \frac{1}{r!} \sum_{|I|=r} c_I\, dt_I \in \bigwedge^r \mathbf{C}\{dt_1, \cdots, dt_{n+q}\}$$

so that

$$H^r(X, \mathbf{C}) \simeq \bigwedge^r \mathbf{C}\{dt_1, \cdots, dt_{n+q}\}.$$

Let $z_j = u_j + iu_{q+j}$ $(j = 1, \cdots, q)$, $z_{q+k} = u_{2q+k} + iv_k$ $(k = 1, \cdots, n-q)$. Then $\binom{u}{v} = A^{-1}t$, $\mathbf{C}\{dt_1, \cdots, dt_{n+q}\} = \mathbf{C}\{du_1, \cdots, du_{n+q}\}$, $v_1 = t_{n+q+1}, \cdots, v_{n-q} = t_{2n}$.
Since v_k are $\mathcal{C}^\infty$–functions on X, $dv_k \sim 0$ such that $dz_{q+k} \sim d\overline{z}_{q+k}$ and $du_{2q+k} \sim dz_{q+k}$. Therefore $dt_1, \cdots, dt_{n+q}$ are d–cohomologous to linear compositions of $dz_1, \cdots, dz_n, d\overline{z}_1, \cdots, d\overline{z}_q$ and vice versa. We have a 1-1-mapping

$$\bigwedge^r \mathbf{C}\{dt_1, \cdots, dt_{n+q}\} \ni \chi \longrightarrow \chi_{\mathbf{C}} \in \bigwedge^r \mathbf{C}\{dz_1, \cdots, dz_n, d\overline{z}_1, \cdots, d\overline{z}_q\},$$

where

$$\chi_{\mathbf{C}} = \sum_{m+p=r} \frac{1}{m!p!} \sum_{\substack{1\le j_1,\cdots,j_m \le n \\ 1\le k_1,\cdots,k_p \le q}} c'_{j_1\cdots j_m \overline{k}_1\cdots \overline{k}_p}\, dz_{j_1} \wedge \cdots \wedge dz_{j_m} \wedge d\overline{z}_{k_1} \wedge \cdots \wedge d\overline{z}_{k_p}.$$

Thus we get the following Theorem about de Rahm cohomology due to UMENO in 1993 [113].

3.2.5 Theorem (UMENO)
Let $X = \mathbf{C}^n/\Lambda$ be a toroidal group of type q.
Then the following statements hold in toroidal coordinates:
a) The cohomology class of $\phi \in Z(X, \mathcal{E}^r)$ is represented by uniquely determined

$$\chi \in \bigwedge^r \mathbf{C}\{dt_1, \cdots, dt_{n+q}\}$$

and

$$\chi_{\mathbf{C}} \in \bigwedge^r \mathbf{C}\{dz_1, \cdots, dz_n, d\overline{z}_1, \cdots, d\overline{z}_q\}.$$

b)
$$H^r(X, \mathbf{C}) \simeq \bigwedge^r \mathbf{C}\{dt_1, \cdots, dt_{n+q}\},$$
$$\simeq \bigwedge^r \mathbf{C}\{dz_1, \cdots, dz_n, d\overline{z}_1, \cdots, d\overline{z}_q\} \ (1 \leq r \leq n+q).$$

$$H^r(X, \mathbf{C}) = 0 \qquad (r > n+q),$$

then

$$\dim H^r(X, \mathbf{C}) = \binom{n+q}{r} \quad (1 \leq r).$$

For toroidal theta groups $X = \mathbf{C}^n/\Lambda$ we got after Theorem 2.2.6

$$H^p(X, \Omega^m) \simeq \bigwedge^m \mathbf{C}\{dz_1, \cdots, dz_n\} \wedge \bigwedge^p \mathbf{C}\{d\overline{z}_1, \cdots, d\overline{z}_q\}.$$

The consequence is the following Theorem of VOGT in 1983 [117] (which is obviously wrong for toroidal wild groups).

3.2.6 Theorem (VOGT)
For toroidal *theta* groups $X = \mathbf{C}^n/\Lambda$ of type q

$$H^r(X, \mathbf{C}) \simeq \bigoplus_{\substack{m+p=r \\ m \leq n, p \leq q}} H^p(X, \Omega^m). \qquad \textbf{(Hodge decomposition)}$$

Proof
As we have seen before any r–form $\chi_{\mathbf{C}}$ has a decomposition

$$\chi_{\mathbf{C}} = \sum_{\substack{m+p=r \\ m \leq n, p \leq q}} \chi^{m,p} \quad \text{with} \quad \chi^{m,p} := \frac{1}{m!}\frac{1}{p!} \sum_{\substack{|J|=m \\ |K|=p}} c'_{J\overline{K}} dz_J \wedge d\overline{z}_K$$

so that the mappings

$$\Phi^{m,p} : H^r(X, \mathbf{C}) \ni [\chi] \to [\chi^{m,p}] \in H^p(X, \Omega^m)$$

induce a homomorphism

$$\Phi : H^r(X, \mathbf{C}) \longrightarrow \bigoplus_{\substack{m+p=r \\ m \leq n, p \leq q}} H^p(X, \Omega^m).$$

Considering the dimensions, we see that Φ is an isomorphism. *Q.E.D.*

Chern forms of fibre metrics

Let X be a complex manifold of dimension n.

A **Hermitian metric** on X is given in local coordinates $z_1, \cdots, z_n$ by

$$ds^2 = \sum_{j,k=1}^{n} h_{j\overline{k}}(x)\, dz_j d\overline{z}_k$$

where the $\mathcal{C}^\infty$–coefficient matrix $(h_{j\overline{k}})_{j,k=1,\cdots,n}$ is Hermitian and positive definite on X.

Every complex manifold X has a Hermitian metric [76, p 83].

A **Hermitian form** $\omega = \sum_{j,k=1}^{n} h_{j\overline{k}}\, dz_j \wedge d\overline{z}_k$ on X is a $\mathcal{C}^\infty$-form $\omega \in \mathcal{E}^{1,1}(X)$ with a Hermitian coefficient matrix $H = (h_{j\overline{k}})_{j,k=1,\cdots,n}$.

If ω is a Hermitian form, then $i\omega$ is a real form.

Let $L \to X$ be a holomorphic line bundle on X which is defined by a 1-cocycle $\{g_{\nu\mu}\}$ belonging to an open covering $\{U_\nu\}_{\nu \in N}$ of X.

A **(Hermitian) fibre metric** h on L is given by a set of $\mathcal{C}^\infty$-functions

$$(h) \qquad h_\nu : U_\nu \to \mathbf{R}_{>0} \quad \text{with} \quad h_\mu = h_\nu\, |g_{\nu\mu}|^2 \qquad (\nu, \mu \in N).$$

Let ζ_ν be fibre coordinates on U_ν $(\nu \in N)$. The induced Hermitian product $\langle \zeta_\nu, \zeta_\nu \rangle := h_\nu |\zeta_\nu|^2$ remains invariant against coordinate transformations, because $h_\nu \zeta_\nu \overline{\zeta}_\nu = h_\nu g_{\nu\mu} \zeta_\mu \overline{g}_{\nu\mu} \overline{\zeta}_\mu = h_\mu \zeta_\mu \overline{\zeta}_\mu \quad (\nu, \mu \in N)$.

Every line bundle on any complex manifold has a Hermitian fibre metric [76, p 100]. A line bundle with a Hermitian fibre metric is a **Hermitian line bundle**.

If L_1, L_2 are line bundles on X with fibre metrics h_1, h_2 resp., then $L_1 \otimes L_2$ has the fibre metric $h_1 h_2$. If L has the fibre metric h, then the dual bundle L^* has the fibre metric $1/h$. If $f : X \to Y$ is a holomorphic map of complex manifolds, then $f \circ h$ is a fibre metric on the pullback f^*L of L.

A Hermitian fibre metric h on L defines on X the Hermitian form

$$(\theta) \quad \theta_h := -\partial\overline{\partial} \log h_\nu$$
$$= -\sum_{j,k=1}^{n} \partial_j \overline{\partial}_k \log h_\nu\, dz_j^\nu \wedge d\overline{z}_k^\nu = -\sum_{j,k=1}^{n} \frac{\partial^2 \log h_\nu}{\partial z_j^\nu \partial \overline{z}_k^\nu} dz_j^\nu \wedge d\overline{z}_k^\nu .$$

This form is globally defined on X and d–closed.

3.2.7 Definition

Let L be a line bundle on X with a fibre metric h.

Then $X_{j\overline{k}} := -\partial_j \overline{\partial}_k \log h$ $(j, k = 1, \cdots, n)$ is called the **curvature**, $\theta_h := -\partial\overline{\partial} \log h$ the **curvature form** and $\Theta_h := -\frac{1}{2\pi i}\theta_h$ the **Chern form** of h.

The inclusion $\mathbf{Z} \hookrightarrow \mathbf{C}$ induces in Čech cohomology the inclusion $H^2(X,\mathbf{Z}) \hookrightarrow H^2(X,\mathbf{C})$ so that the Chern class $c_1(L)$ can be assumed to be in $H^2(X,\mathbf{C})$. By de Rahm $H^2(X,\mathbf{C}) \simeq \{d\text{–closed } \mathcal{C}^\infty \text{ 2–forms on } X\} / \{d\text{–exact } \mathcal{C}^\infty \text{ 2–forms on } X\}$ so that we can consider the d-closed $(1,1)$–forms obviously as part of $H^2(X,\mathbf{C})$.

3.2.8 Proposition
The Chern class $c_1(L)$ of a line bundle L on a complex manifold X in $H^2(X,\mathbf{C})$ is represented by every Chern form

$$(\Theta) \qquad \Theta_h := -\frac{1}{2\pi i}\theta_h = \frac{1}{2\pi i}\partial\bar\partial \log h$$

with any fibre metric h of L.

For the *proof* confer GRIFFITHS and HARRIS [39, p 141 and 148f] or MORROW and KODAIRA [76, p 127]. In the first book $-\theta_h$ is used as curvature form.

We use the following definition of a positive line bundle.

3.2.9 Definition
A line bundle on a complex manifold is **positive**, iff it has a Hermitian fibre metric whose curvature form is positive definite.

Every line bundle L on a toroidal group is defined by an automorphic factor with the Hermitian decomposition 3.1.1. This decomposition determines the Hermitian form H on $\mathbf{C}^n$, however uniquely only on $\mathbf{R}_\Lambda$. We say that the **line bundle** L **determines** H. A line bundle is **ample**, iff the Hermitian form H is positive definite.

We want to establish a connection between the Hermitian forms determined by a line bundle and the averages of the Chern forms of the fibre metrics of the same bundle.

The following lemma is due to ABE (Remark 1 in [2]).

3.2.10 Lemma
Let a line bundle L on a toroidal group $X = \mathbf{C}^n/\Lambda$ be defined by an automorphic factor α_λ $(\lambda \in \Lambda)$ with an exponential system of functions a_λ $(\lambda \in \Lambda)$.
Then the Hermitian fibre metrics h on L correspond to the real valued $\mathcal{C}^\infty$–functions A on $\mathbf{C}^n$ with

$$A(z+\lambda) - A(z) = 4\pi \operatorname{Im} a_\lambda(z) \qquad (\lambda \in \Lambda),$$

which satisfies

$$\left(\frac{\partial^2 \log h}{\partial z_j \partial \overline{z_k}}\right)_{j,k=1,\cdots,n} = \left(\frac{\partial^2 A}{\partial z_j \partial \overline{z_k}}\right)_{j,k=1,\cdots,n}.$$

Proof

$\succ$. First we note that a consequence of the cocycle property of exponential systems is (see p 28)

$$(*) \qquad \operatorname{Im} a_{\lambda_1+\lambda_2}(z) = \operatorname{Im} a_{\lambda_2}(z+\lambda_1) + \operatorname{Im} a_{\lambda_1}(z) \qquad (\lambda_1, \lambda_2 \in \Lambda).$$

The natural projection $\pi : \mathbf{C}^n \to \mathbf{C}^n/\Lambda$ is a covering map. there exists a finite covering $(U_\nu)_{\nu \in N}$ of X so that all $\pi^{-1}(U_\nu) = \bigcup_{\lambda \in \Lambda}\{U_\nu^* + \lambda\}$ where U_ν^* is one of the pairwise disjoint countable many components of $\pi^{-1}(U_\nu)$ biholomorphic to U_ν by π. We can assume $U_\mu^* \cap U_\nu^* = \emptyset$ $(\mu \neq \nu)$ and set $\varrho_\nu := (\pi|_{U_\nu^*})^{-1}$. If $U_\nu \cap U_\mu \neq \emptyset$, then there exists a $\lambda_{\nu\mu} \in \Lambda$ with $U_\nu^* \cap (U_\mu^* + \lambda_{\nu\mu}) \neq \emptyset$.
Now the transition functions defining L can be described on X as

$$g_{\nu\mu}(x) = \alpha_{\lambda_{\nu\mu}}(\varrho_\mu(x)) \qquad (x \in U_\nu \cap U_\mu).$$

By definition (h) of the fibre metric on L (p 80) we have

$$2 \log|g_{\nu\mu}(x)| = \log(h_\mu(x)) - \log(h_\nu(x)) \qquad (x \in U_\nu \cap U_\mu).$$

We define on $V := \bigcup U_\nu^*$ the C^∞-function

$$A^*(z) := \log(h_\nu(\pi(z))) \qquad (z \in U_\nu^*).$$

If $z \in U_\nu^*$ and $z + \lambda_{\mu\nu} \in U_\mu^*$, then

$$A^*(z+\lambda_{\mu\nu}) - A^*(z) = 2\log|g_{\nu\mu}(\pi(z))| = -4\pi \operatorname{Im} a_{\lambda_{\nu\mu}+\lambda_{\mu\nu}}(z) = 4\pi \operatorname{Im} a_{\lambda_{\mu\nu}}(z).$$

For any $z \in \mathbf{C}^n$ there exists a $\lambda \in \Lambda$ with $z + \lambda \in U_\nu^*$ for a suitable ν so that

$$A(z) := A^*(z+\lambda) + 4\pi \operatorname{Im} a_\lambda(z) \qquad (z \in \mathbf{C}^n)$$

is well defined by the help of $(*)$.
Finally we need $(*)$ once more to see the desired property of A.
$\prec$. Directly with $h_\nu := \exp(A \circ \varrho_\nu)$ $(\nu \in N)$ and just this fibre metric fulfils the additional condition. *Q.E.D.*

Abe [6] proved in 1989 that every theta bundle on a toroidal group determines a Hermitian form H on $\mathbf{C}^n$ which is given by a certain fibre metric on this bundle.

3.2.11 Theorem (Abe)

Let $X = \mathbf{C}^n/\Lambda$ be a toroidal group and L a theta bundle on X determining the Hermitian form H. Then there exists a fibre metric h on L such that

$$\pi\Big(H_{j\bar{k}}\Big)_{j,k=1,\cdots,n} = \Big(-\frac{\partial^2 \log h}{\partial z_j \partial \overline{z}_k}\Big)_{j,k=1,\cdots,n}.$$

Proof
The Appell–Humbert decomposition of an automorphic factor defining the theta bundle L is

$$\vartheta_\lambda(z) = \mathbf{e}(a_\lambda(z)) = \varrho(\lambda)\mathbf{e}\Big(\frac{1}{2i}\Big[H(z,\lambda) + \frac{1}{2}H(\lambda,\lambda)\Big]\Big) \qquad (\lambda \in \Lambda)$$

with a Hermitian form H and semi–character $\varrho : \Lambda \to \mathbf{S}^1$ so that

$$4\pi \operatorname{Im} a_\lambda(z) = -(2\pi \operatorname{Re} H(z,\lambda) + \pi H(\lambda,\lambda)).$$

Let $A(z) := -\pi \operatorname{Re} H(z,z)$. Then we calculate $A(z+\lambda) - A(z) = 4\pi \operatorname{Im} a_\lambda(z)$ and $-\partial^2 A/\partial z_j \partial \overline{z_k} = \pi H_{j\overline{k}}$. Using the foregoing lemma we get

$$\pi\Big(H_{j\overline{k}}\Big)_{j,k=1,\cdots,n} = \Big(-\frac{\partial^2 \log h}{\partial z_j \partial \overline{z_k}}\Big)_{j,k=1,\cdots,n}$$

for the fibre metric h on L corresponding to $A(z)$. *Q.E.D.*

By Lemma 3.2.3(4) the average of a Chern form represents the same Chern class $c_1(L)$ as the Chern form itself. This average has by definition an associated form which is Hermitian.

For the last theorem we need a Lemma of ABE [10].

3.2.12 Lemma (ABE)
Let L be a line bundle on a toroidal group $X = \mathbf{C}^n/\Lambda$. Then:

1. If L is topologically trivial, then

$$Av(\Theta_h) = 0 \text{ on } T_K \times T_K$$

for any Hermitian fibre metric h on L.
2. The average $Av(\Theta_h)$ of the Chern form of a Hermitian fibre metric h on L is on $T_K \times T_K$ independent of h.

Proof
1. If L is topologically trivial, then $c_1(L) = 0$ by Proposition 2.1.9 on p 38. Then the Chern form Θ_h of any fibre metric is d–exact. By Lemma 3.2.3(2) we have

$$Av(\Theta_h) = 0 \text{ on } T_K \times T_K.$$

2. Let h_1, h_2 be two fibre metrics on the given line bundle. Then $h := h_1/h_2$ is a fibre metric on the trivial line bundle. Applying 1) to the Chern form Θ_h we obtain the conclusion. *Q.E.D.*

Let $\Theta_h = \frac{1}{2\pi i}\partial\overline{\partial}\log h$ be the Chern form of a fibre metric h on a line bundle L. Then its average

$$\tilde{\Theta}_h := Av(\Theta_h) = -\frac{1}{2\pi i}\sum_{j,k=1}^{n} \tilde{h}_{j\bar{k}} dz_j \wedge d\bar{z}_k$$

is by Lemma 3.2.12 on $T_K \times T_K$ independent of h.

We state this meaning more precisely. Let $z_j = u_j + iu_{q+j}$ $(j = 1, \cdots, q)$ and $z_{q+k} = u_{2q+k} + iv_k$ $(k = 1, \cdots, n-q)$ in toroidal coordinates.
The relation between real coordinates (u, v) and $t = (t', t'')$ is $\binom{u}{v} = A^{-1}t$, especially $v_1 = t_{n+q+1}, \cdots, v_{n-q} = t_{2n}$. Then, in the representation of $\tilde{\Theta}_h$, the coefficients $\tilde{h}_{j\bar{k}}$ are dependent only on $v = t''$

$$\tilde{h}_{j\bar{k}}(v) = \tilde{h}_{j\bar{k}}(t'') = \int_K \left(-\frac{\partial^2 \log h}{\partial z_j \partial \bar{z}_k}(t', t'')\right) dt'.$$

Since T_K is generated by $\partial/\partial t_1, \cdots, \partial/\partial t_{n+q}$ and also by $\partial/\partial u_1, \cdots, \partial/\partial u_{n+q}$, $\tilde{\Theta}_h$ has the h-invariant part consisting only of terms $du_i \wedge du_j$, if we rewrite $\tilde{\Theta}_h$ in terms $du_1, \cdots, du_{n+q}, dv_1, \cdots, dv_{n-q}$. The average $\tilde{\Theta}_h$ gives the associated Hermitian form $\tilde{H}^h_v(z, w)$ on $\mathbf{C}^n$ defined by

$$\tilde{H}^h_v(z, w) = \pi \sum_{j,k=1}^{n} \tilde{h}_{j\bar{k}}(v) z_j \overline{w}_k.$$

The above fact means that $\tilde{H}^h_v|_{\mathbf{R}_\Lambda \times \mathbf{R}_\Lambda}$ is independent of h.
In particular, $\tilde{H}^h_v|_{\mathbf{R}_\Lambda \times \mathbf{R}_\Lambda} = 0$ if L is topologically trivial.

Two Hermitian forms H_1 and H_2 on $\mathbf{C}^n$ are **Λ-equivalent**, iff

$$H_1|_{\mathbf{R}_\Lambda \times \mathbf{R}_\Lambda} = H_2|_{\mathbf{R}_\Lambda \times \mathbf{R}_\Lambda} \quad \text{abbreviated as } H_1 \simeq_\Lambda H_2.$$

The following theorem of ABE [10] characterizes the positive line bundles on toroidal groups.

3.2.13 Theorem (ABE)
Let L be a holomorphic line bundle on a toroidal group $X = \mathbf{C}^n/\Lambda$ determining the Hermitian form $H(z, w) = \sum_{j,k=1}^{n} H_{j\bar{k}} z_j \overline{w}_k$. Then the Hermitian form $\tilde{H}^h_v$ associated with the average of the Chern form Θ_h of any Hermitian fibre metric h on L is Λ-equivalent to H.
If L is positive, then H is an ample Riemann form for Λ.
Proof
Let $L = L_\vartheta \otimes L_0$ be the decomposition of L into a theta bundle L_ϑ and a topologically trivial line bundle L_0 (see Theorem 2.1.10). By Theorem 3.2.11 there exists a fibre metric h_ϑ on L_ϑ so that $\Theta_{h_\vartheta} = \frac{1}{2\pi i}\partial\bar{\partial} \log h_\vartheta = -\frac{1}{2i}\sum_{j,k=1}^{n} H_{j\bar{k}} dz_j \wedge d\bar{z}_k$.
Then the average $\tilde{\Theta}_{h_\vartheta} := Av(\Theta_{h_\vartheta})$ is obviously the same:

$$\tilde{\Theta}_{h_\vartheta} = \Theta_{h_\vartheta} = -\frac{1}{2i} \sum_{j,k=1}^{n} H_{j\bar{k}} dz_j \wedge d\bar{z}_k .$$

Therefore the Hermitian form $\tilde{H}_v^{h_\vartheta}$ associated with $\tilde{\Theta}_{h_\vartheta}$ is just H.
For any fibre metric h_0 on L_0, the Hermitian form $\tilde{H}_v^{h_\vartheta}$ associated with the average of the Chern form Θ_{h_ϑ} vanishes on $\mathbf{R}_\Lambda \times \mathbf{R}_\Lambda$.
Now $h_1 := h_\vartheta h_0$ is a fibre metric on L. The associated form $\tilde{H}_v^{h_1}$ is $\tilde{H}_v^{h_\vartheta} + \tilde{H}_v^{h_0}$. Then

$$\begin{aligned}\tilde{H}_v^{h_1}|_{\mathbf{R}_\Lambda \times \mathbf{R}_\Lambda} &= \tilde{H}_v^{h_\vartheta}|_{\mathbf{R}_\Lambda \times \mathbf{R}_\Lambda} + \tilde{H}_v^{h_0}|_{\mathbf{R}_\Lambda \times \mathbf{R}_\Lambda} \\ &= H.\end{aligned}$$

For any fibre metric h on L, we obtain by Lemma 3.2.12

$$\tilde{H}_v^h \simeq_\Lambda \tilde{H}_v^{h_1} \simeq_\Lambda H.$$

Finally let L be positive. To know which H is an ample Riemann form for Λ, we need only to know that $H > 0$ on MC_Λ. Let h be a fibre metric on L such that Θ_h is a positive form. Then its average $\tilde{\Theta}_h$ is also positive. Since the associated Hermitian form $\tilde{H}_v^h$ and H coincide on $\mathrm{MC}_\Lambda \subset \mathbf{R}_\Lambda$, $H > 0$ on MC_Λ. *Q.E.D.*

Holomorphic mappings to projective spaces

We study holomorphic sections of line bundles which define maps to complex projective spaces.

3.2.14 Definition
Let L be a holomorphic line bundle on $X = \mathbf{C}^n/\Lambda$.
The **space $H^0(X, L)$ of all holomorphic sections of L generates L on X**, iff for any $x \in X$ there exists a section $\varphi \in H^0(X, L)$ such that $\varphi(x) \neq 0$.

The following lemma was proved by Capocasa and Catanese in 1991 [20] and Abe in 1993 [8].
For that let L be given by an automorphic factor α. As we have seen in Proposition 2.1.2 on p 27 $H^0(X, L)$ is isomorphic to the space $\mathcal{A}_\alpha$ of all automorphic forms belonging to α. Then obviously $H^0(X, L^\ell)$ is isomorphic to $\mathcal{A}_{\alpha^\ell}$.

3.2.15 Lemma
Let L be a holomorphic line bundle on $X = \mathbf{C}^n/\Lambda$.
If $H^0(X, L) \neq 0$, then $H^0(X, L^\ell)$ generates L^ℓ on X for any integer $\ell \geq 2$.
Proof
For $l = 2$. (The general proof works in the same way.)

Because $H^0(X,L) \neq 0$ there exists an automorphic form $f \not\equiv 0$. Then let us consider $F(a,z) := f(z-a)f(z+a)$ $(a, z \in \mathbf{C}^n)$ and $f_0(a) := F(a, z^{(0)})$ $(a \in \mathbf{C}^n)$ for any fixed $z^{(0)} \in \mathbf{C}^n$. f_0 cannot be $\equiv 0$ on MC_Λ. Otherwise, because $\mathrm{MC}_\Lambda + \Lambda$ is dense in $\mathbf{R}_\Lambda$, f would be even $\equiv 0$ on $\mathbf{R}_\Lambda$, and then on $\mathbf{C}^n$. So there exists an $a^{(0)} \in \mathrm{MC}_\Lambda$ with $f_0(a^{(0)}) \neq 0$.
Now let $\alpha_\lambda = \vartheta_\lambda \, \iota_\lambda$ $(\lambda \in \Lambda)$ be the automorphic factor defining $L = L_\vartheta \otimes L_0$ where the theta factor ϑ_λ defines the theta bundle L_ϑ and the wild factor ι_λ the topologically trivial bundle L_0 as described in Theorem 3.1.4.
Then $F(a^{(0)}, z)$ $(z \in \mathbf{C}^n)$ becomes an automorphic form for

$$\vartheta_\lambda(z-a^{(0)})\,\vartheta_\lambda(z+a^{(0)})\,\iota_\lambda(z-a^{(0)})\,\iota_\lambda(z+a^{(0)}).$$

The product of the first two reduced theta factors of type (H, ϱ) is the reduced theta factor ϑ_λ^2 of type $(2H, \varrho^2)$ and the product of the last two wild factors is the wild factor $\iota_\lambda^2(z)$ for $a^{(0)} \in \mathrm{MC}_\Lambda$. So $F(a^{(0)}, z)$ is an automorphic form of $\mathcal{A}_{\alpha^2}$ and $F(a^{(0)}, z^{(0)}) = f_0(a^{(0)}) \neq 0$. *Q.E.D.*

Remark
Taking suitable $a^{(0)}$'s as in the proof of the above lemma, we get relatively prime sections φ and ψ of L^ℓ.

$H^0(X,L)^N$ is a complete metric space.
Indeed, let $\{V_j\}$ and $\{W_j\}$ with $V_j \Subset W_j \Subset X$ $(j \in J)$ be locally finite coverings of X, such that L is trivial on every W_j. Then

$$|\varphi|_C := \max_j \sup_{C \cap V_j} |\varphi| \quad (C \subset X \text{ compact})$$

define semi-norms on $H^0(X,L)$, by which $H^0(X,L)$ becomes a Fréchet space, therefore a complete metric space and then also $H^0(X,L)^N$.
We say that $\Phi = (\varphi^{(1)}, \cdots, \varphi^{(N)}) \in H^0(X,L)^N$ *has zeros* iff the sections $\varphi^{(1)}, \cdots, \varphi^{(N)}$ have common zeros.

The following proposition of Abe guarantees the existence of holomorphic mappings to higher dimensional projective spaces.

For that we remember that the image of a compact set of a complex manifold X of dimension n under a holomorphic map $F = (f_1, \cdots, f_N) : X \to \mathbf{C}^N$ is a compact Lebesgue zero set, if $N > n$ [46, p 114].

3.2.16 Proposition (Abe)
Suppose that $H^0(X,L)$ generates L on $X = \mathbf{C}^n/\Lambda$ and let $N \geq n$. Then the set of $(N+1)$-tuples of sections $\Phi \in H^0(X,L)^{N+1}$ with zeros on X is of first category.
Proof
It is sufficient to prove the statement for any compact subset $C \subset X$, because

X is the countable union of compact sets. So we consider

$$M := \{\Phi \in H^0(X,L)^{N+1} : \Phi \text{ has a zero on } C\}.$$

Statement a. M is closed in $H^0(X,L)^{N+1}$: Let $M \ni \Phi_j \to \Phi$ in $H^0(X,L)^{N+1}$. For each j take $x_j \in C$, such that $\Phi_j(x_j) = 0$. We can assume that $x_j \to x_0 \in C$. Because Φ_j converges uniformly on C we get $0 = \Phi_j(x_j) \to \Phi(x_0) = 0$ so that $\Phi \in M$.

Statement b. For any open $U \subset\subset X$ there exists a finite number of sections $\psi^{(0)}, \cdots, \psi^{(r)} \in H^0(X,L)$ without any common zero in U:

Indeed, without common zero even in compact $\overline{U}$ by assumption.

Statement c. Let $(\varphi^{(0)}, \cdots, \varphi^{(N)}, \psi) \in H^0(X,L)^{N+2}$ without zeros in C and $N \geq n$. Then there exists a Lebesgue zero set $Z \subset \mathbf{C}^{N+1}$ such that

$$(\varphi^{(0)} - a_0\psi, \cdots, \varphi^{(N)} - a_N\psi) \in H^0(X,L)^{N+1}$$

has no zeros on C for any $(a_1, \cdots, a_N) \in \mathbf{C}^{N+1} \setminus Z$: For that we can assume without loss of generality that the compact set C is contained in some coordinate neighborhood U on which L is trivial. The sections $\varphi^{(0)}, \cdots, \varphi^{(N)}, \psi$ can be considered to be holomorphic functions on U. Let $U_0 := \{x \in U : \psi(x) \neq 0\}$. If $U_0 \cap C = \emptyset$ we are ready with $Z := \emptyset$. If $U_0 \cap C \neq \emptyset$, we set $U_0 \cap C =: \bigcup_{j=1}^{\infty} C_j$ with compact C_j. Because $\Phi_\psi := (\phi^{(0)}/\psi, \cdots, \phi^{(N)}/\psi) : U_0 \to \mathbf{C}^{N+1}$ is a holomorphic map, all $\Phi_\psi(C_j)$ are compact sets of Lebesgue measure zero in $\mathbf{C}^{N+1}$ according to the remark before formulating this proposition. Then we can take $Z := \Phi_\psi(U_0 \cap C)$.

Now we can finish the proof of the proposition. The rest is to show that M has no interior point. Let $\Phi \in M$. After statement b) there exists a $\Psi = (\psi^{(0)}, \cdots, \psi^{(r)}) \in H^0(X,L)^{r+1}$ without zeros on C. Applying statement c) successivly to $(\Phi, \Psi) \in H^0(X,L)^{N+r+2}$ we can conclude that

$$(\varphi^{(0)} - \sum_{j=0}^{r} a_{0j}\psi^{(j)}, \cdots, \varphi^{(N)} - \sum_{j=0}^{r} a_{Nj}\psi^{(j)}) \in H^0(X,L)^{N+1}$$

has no zeros on C for suitable and arbitrarily small coefficients a_{ij}. So Φ is no interior point of M. *Q.E.D.*

3.2.17 Theorem (ABE)

Suppose that $H^0(X,L)$ generates L on X and $N > n$. Then there exists a holomorphic mapping

$$\Phi = (\varphi^{(0)} : \cdots : \varphi^{(N)}) : X \longrightarrow \mathbf{P}_N \text{ with } \varphi^{(j)} \in H^0(X,L) \text{ and } L = \Phi^*[e],$$

where $[e]$ is the hyperplane bundle of the N-dimensional projective space $\mathbf{P}_N$.

Proof

By the previous proposition there exist sections $\varphi^{(0)}, \cdots, \varphi^{(N)} \in H^0(X,L)$ hav-

ing no common zero on X. They define a map $\Phi : X \longrightarrow \mathbf{P}_{N+1}$.
The fact that $L = \Phi^*[e]$ is trivial. *Q.E.D.*

Meromorphic functions on toroidal groups

KOPFERMANN proved in 1960 for torus groups and in 1964 for toroidal groups, that the Hermitian form determined by a theta factor is positiv semi–definite, if the theta factor has a non–trivial automorphic form [63, 64].
ABE generalized in 1995 this property for *all* line bundles on toroidal groups and proved that the meromorphic functions are constant on the fibres of the Hermitian form which is determined by the given line bundle [10].

The additive group of divisors can be identified by

$$(*) \qquad \mathrm{Div}(X) = H^0(X, \mathcal{M}^*/\mathcal{O}^*),$$

where $\mathcal{M}^*$ is the multiplicative sheaf of germs of meromorphic functions $\not\equiv 0$ and $\mathcal{O}^*$ the subsheaf of germs of holomorphic functions $\not\equiv 0$.
A divisor D given by local functions $f_\mu \in \mathcal{M}^*(U_\mu)$ on an open covering $\{U_\mu\}$ of X defines the transition functions $g_{\mu\nu} := f_\mu / f_\nu$ of a line bundle $[D]$. If D, D' are divisors, then $D + D'$ defines $[D] \otimes [D']$ so that the map $D \xrightarrow{\delta} [D]$ becomes a homomorphism which can be described by a cohomology sequence.
Indeed, the exact sequence

$$0 \to \mathcal{O}^* \xrightarrow{\iota} \mathcal{M}^* \xrightarrow{p} \mathcal{M}^*/\mathcal{O}^* \to 0$$

induces the cohomology sequence

$$\cdots \to H^0(X, \mathcal{M}^*) \xrightarrow{p_*} H^0(X, \mathcal{M}^*/\mathcal{O}^*) \xrightarrow{\delta} H^1(X, \mathcal{O}^*) \to \cdots .$$

Using $(*)$ and the definition of the Picard group we get the same homomorphism

$$\delta : \mathrm{Div}(X) \to \mathrm{Pic}(X).$$

Let $X = \mathbf{C}^n/\Lambda$ be a toroidal group and H a Hermitian form on $\mathbf{C}^n$ with imaginary part $A := \operatorname{Im} H$. Then we write

$$H_\Lambda := H|_{\mathbf{MC}_\Lambda \times \mathbf{MC}_\Lambda} \text{ and } A_\Lambda := A|_{\mathbf{R}_\Lambda \times \mathbf{R}_\Lambda}.$$

3.2.18 Theorem (ABE)
Let L be a holomorphic line bundle over a toroidal group $X = \mathbf{C}^n/\Lambda$ which determines the Hermitian form H on $\mathbf{C}^n$.
Suppose that $H^0(X, L)$ generates L on X or suppose $H^0(X, L) \neq 0$. Then:

1. there exists a positive semi-definite Hermitian form $\tilde{H}$ on the $\mathbf{C}^n$ which is Λ-equivalent to H.
2. For any $\psi_1, \psi_2 \in H^0(X,L)$ with $\psi_2 \not\equiv 0$ the meromorphic function $f := \psi_1/\psi_2$ is constant[1] on $x_0 + \pi(\mathrm{Ker}H_\Lambda)$ for all $x_0 \in X$, where $\pi : \mathbf{C}^n \to X = \mathbf{C}^n/\Lambda$ is the projection.

Proof

a) If $H^0(X,L)$ generates L, then by Theorem 3.2.17 there exists a holomorphic mapping

$$\Phi = (\varphi^{(0)} : \cdots : \varphi^{(N)}) : X \longrightarrow \mathbf{P}_N \text{ with } \varphi^{(j)} \in H^0(X,L) \text{ and } L = \Phi^*[e].$$

Since $(\psi_1 : \psi_2 : \varphi^{(0)} : \cdots : \varphi^{(N)}) : X \longrightarrow \mathbf{P}_{N+2}$ has the same property we can assume $\varphi^{(0)} = \psi_1, \varphi^{(1)} = \psi_2$.

The hyperplane bundle $[e]$ on $\mathbf{P}_n$, defined by the hyperplane divisor $e = \mathbf{P}_{n-1} \subset \mathbf{P}_n$, is positive so that there exists a fibre metric h with a positive curvature form.[2] With the fibre metric h we define

$$E := \frac{1}{\pi}\Big(-\frac{\partial^2 \log h}{\partial x_j \partial \overline{x}_k}\Big)_{j,k=1,\cdots,N}$$

and the pullback $h \circ \Phi$ induces the Hermitian matrix

$$\tilde{E} = (\tilde{E}_{j\overline{k}}(z)) := \frac{1}{\pi}\Big(-\frac{\partial^2 \log(h \circ \Phi)}{\partial z_j \partial \overline{z}_k}\Big)_{j,k=1,\cdots,n}.$$

We have $\tilde{E} = {}^t\partial\Phi\, E\, \overline{\partial\Phi}$, where $\partial\Phi$ is the Jacobian of Φ and by Theorem 3.2.13 for the Hermitian form $\tilde{H}_v$ given by $(\int_K \tilde{E}_{j\overline{k}}(t',v)dt')_{j,k=1,\cdots,n}$ is Λ-equivalent to H. Since $\tilde{E}$ is positive semi-definite on $\mathbf{C}^n$, $\tilde{H}_v$ is also positive semi-definite. For any $\zeta \in \mathrm{Ker}\, H_\Lambda \subset \mathbf{MC}_\Lambda$ we get

$$0 = H(\zeta,\zeta) = \tilde{H}_v(\zeta,\zeta) = \int_K {}^t[(\partial\Phi)\zeta]\, E\, \overline{(\partial\Phi)\zeta} dt'.$$

Then in a neighborhood of a point x_0 we have $(\partial\Phi)\zeta = 0$ for E is positive definite. Therefore Φ and then f is constant on $x_0 + \pi(\mathrm{Ker}H_\Lambda)$.

b) By Lemma 3.2.15 L^2 satisfies the assumption of a) if $H^0(X,L) \neq 0$. Besides L^2 determines the Hermitian form $2H$. For $2H$ there exists a positive semi-definite Hermitian form $\tilde{H}$ as in a), such that $\tilde{H} \simeq_\Lambda 2H$. The rest of the proof is the same as in a) such that f^2 and then f is constant on $x_0 + \pi(\mathrm{Ker}H_\Lambda)$. *Q.E.D.*

The following two theorems are the last decisive steps to the Main Theorem.

[1] Note that a meromorphic function is called even constant on every set contained in the set of its indeterminacy.

[2] The Chern form of the Hermitian fibre metric h on $[e]$ is given by the associated $(1,1)$-form of the Fubini-Study metric on $\mathbf{P}_n$.

3.2.19 Theorem (ABE)
Any meromorphically separable toroidal group is a quasi-Abelian variety.
Proof
Let $X = \mathbf{C}^n/\Lambda$ be a meromorphically separable toroidal group. Define $\mathcal{T}$ as the set of all theta factors ϑ with a type determining Hermitian forms H_ϑ, which are positive semi-definite on MC_Λ, and let

$$\operatorname{Ker}\mathcal{T} := \bigcap_{\vartheta\in\mathcal{T}} \operatorname{Ker}(H_\vartheta)_\Lambda.$$

It suffices to show that $\operatorname{Ker}\mathcal{T} = 0$.
Indeed, if $\dim_{\mathbf{C}} \operatorname{Ker}\mathcal{T} > 0$, then take $x, y \in \operatorname{Ker}\mathcal{T}$ with $\pi(x) \neq \pi(y)$, where $\pi : \mathbf{C}^n \to X = \mathbf{C}^n/\Lambda$ is the projection. Since X is meromorphically separable, there exists a meromorphic function f on X such that $f(\pi(x)) \neq f(\pi(y))$. Let

$$L := [(f)_0] = [(f)_\infty]$$

be the line bundle determined by the zero divisor of f, and simultaneously by the pole divisor of f. We may assume by Theorem 2.1.10 that $L = L_\vartheta \otimes L_0$, where ϑ is a theta factor of a type which determines the Hermitian form H. Now there exist non-zero sections $\varphi, \psi \in H^0(X, L)$ such that $f = \varphi/\psi$. By Theorem 3.2.18 we get $H|_\Lambda \geq 0$ such that $\vartheta \in \mathcal{T}$. By the same Theorem f is constant on the projection of $\operatorname{Ker} H_\Lambda$ and then on the projection of $\operatorname{Ker}\mathcal{T}$. But this contradicts the choice of f. *Q.E.D.*

3.2.20 Theorem (CAPOCASA–CATANESE)
Any toroidal group with a non-degenerate meromorphic function is a quasi-Abelian variety.
Proof
Let f be a non-degenerate Λ-periodic meromorphic function on $\mathbf{C}^n$, where $X = \mathbf{C}^n/\Lambda$ is a toroidal group. Then there exist a holomorphic line bundle L on X and relatively prime holomorphic sections φ and ψ of L such that $f = (\varphi/\psi) \circ \pi$ with the natural projection $\pi : \mathbf{C}^n \to X$. The Hermitian form H determined by the line bundle L is positive semi-definite by Theorem 3.2.18.
If $\operatorname{Ker} H_\Lambda \neq 0$, then f is constant on $z + \operatorname{Ker} H_\Lambda$ for all $z \in \mathbf{C}^n$ by the same theorem. This is not possible because f is non-degenerate. Hence H_Λ is positive definite and therefore an ample Riemann form for Λ. *Q.E.D.*

ABE proved the main results of the Main Theorem for toroidal theta groups in 1987 [1] and for all toroidal groups in 1989 [4] with supplements in [10]. As predecessor GHERARDELLI and ANDREOTTI contributed their Fibration Theorem in 1974 [33]. CAPOCASA and CATANESE added the classical aspect of the existence of non-degenerate meromorphic functions in 1991 [20].

3.2.21 Main Theorem (Characterization of quasi–Abelian varieties)
For a toroidal group $X = \mathbf{C}^n/\Lambda$ the following statements are equivalent:

1.) X is a quasi–Abelian variety.
2.) X has a positive line bundle.
3.) X is the covering group of an Abelian variety.
4.) X has a closed Stein subgroup $N \simeq \mathbf{C}^\ell \times \mathbf{C}^{*m}$ so that X/N is an Abelian variety.
5.) X is quasi–projective.
6.) X is meromorphically separable.
7.) X has a non–degenerate meromorphic function.

Proof
$1 \succ 2$. Theorem 3.2.11.
$2 \succ 1$. Theorem 3.2.13.
$1 \succ 3$. Theorem 3.1.10.
$1 \succ 4$. Fibration Theorem 3.1.16.
$4 \succ 5$. Theorem 3.1.18.
$5 \succ 6$. trivial.
$3 \succ 7$. trivial.
$6 \succ 1$. Theorem 3.2.19.
$7 \succ 1$. Theorem 3.2.20. *Q.E.D.*

In some previous propositions we used the assumption $H^0(X, L) \neq 0$. Cousin showed in 1910 [26] in the special case of dimension 2 and lattices of rank 3, which - in modern language - a topologically trivial line bundle has a non–trivial section, iff it is analytically trivial. Huckleberry and Margulis [47] used his idea and a new one in 1983 to prove the proposition for any dimension n and rank $n + 1$.

The assumption that L is topologically trivial is not necessary.

Abe proved in 1995 [10] the following more general result:

3.2.22 Proposition (Abe)
Let L be a holomorphic line bundle on a toroidal group $X = \mathbf{C}^n/\Lambda$ which determines a Hermitian form H on $\mathbf{C}^n$ and suppose $H_\Lambda = 0$.
Then $H^0(X, L) \neq 0$, iff L is analytically trivial.
Proof
If $H^0(X, L) \neq 0$, then take any $\varphi \in H^0(X, L)$ with $\varphi \not\equiv 0$. By Lemma 3.2.15 $H^0(X, L^2)$ generates L^2 on X and let $2H$ be the Hermitian form determined by L^2. By Theorem 3.2.17 we have a holomorphic mapping

$$\Phi : (\varphi^{(0)} : \cdots : \varphi^{(N)}) : X \to \mathbf{P}_N \quad \text{with} \quad \varphi^{(j)} \in H^0(X, L^2)$$

for which we can suppose $\varphi^{(0)} = \varphi^2$.
Assume now that there exists a zero $x_0 \in X$ of φ. As shown in the proof of Theorem 3.2.18 Φ is constant on the projection of $x_0 + \mathrm{Ker}H_\Lambda = x_0 + \overline{\mathrm{MC}_\Lambda}$ so that $\Phi(x_0 + y) = \Phi(x_0)$ for all $y \in \pi(\mathrm{MC}_\Lambda)$. Therefore $\varphi \equiv 0$ on $x_0 + \pi(\mathrm{MC}_\Lambda)$ and then on X. This is not possible.
The converse is trivial. *Q.E.D.*

The following example of ABE in 1989 [6] shows the existence of non topologically trivial theta bundles with $H_\Lambda = 0$, then $H^0(X, L) = 0$.

Example: The basis in toroidal coordinates

$$P = \begin{pmatrix} 0 & 1 & 0 & i\sqrt{2} & i\sqrt{3} \\ 0 & 0 & 1 & i\sqrt{3} & 0 \\ 1 & \sqrt{2} & 0 & 0 & 0 \end{pmatrix}$$

generates a toroidal group $X = \mathbf{C}^3/\Lambda$. Let $\{e_1, e_2, e_3\}$ be the natural complex basis of $\mathbf{C}^3$. Then $\mathbf{R}_\Lambda$ is generated by $\{e_1, e_2, e_3, ie_1, ie_2\}$. We define an alternating $\mathbf{R}$-bilinear form A on $\mathbf{C}^3 \simeq \mathbf{R}^6$ by

$$\begin{aligned} A(e_1, e_3) &= -A(e_3, e_1) = 1, \\ A(ie_1, ie_3) &= A(e_1, e_3), \\ A(x, y) &= 0 \quad \text{otherwise.} \end{aligned}$$

By A we get a Hermitian form H on $\mathbf{C}^3$ defined by

$$H(x, y) = A(x, iy) + iA(x, y), \quad x, y \in \mathbf{C}^3.$$

Consider the theta factor ϑ whose Hermitian form is H, and the corresponding theta bundle L. Then $H_\Lambda = 0$, but L is not topologically trivial for $A_\Lambda \neq 0$.
In general it is not easy to discuss the conditions for $H^0(X, L) \neq 0$. We shall discuss this problem in the next chapter.

4. Reduction and Extension

ABE showed that the question of the existence of non-trivial sections in line bundles can be reduced to positive line bundles. With a general concept of a Riemann form he gave a new proof of the existence of the meromorphic reduction of a toroidal group. Recently TAKAYAMA proved the conjecture that positive line bundles L always have non-trivial sections. He proved also that L^ℓ is very ample for any integer $\ell \geq 3$. For some questions it is useful to extend holomorphic line bundles from a toroidal group to standard compactifications. M. STEIN showed that this is possible, if and only if the Hermitian form defined by the bundle fulfils a certain condition. ABE condersidered the case where the fibration is associated with an ample Riemann form of kind ℓ.

4.1 Automorphic forms

For the existence of automorphic forms five conditions are necessary. If a given automorphic factor satisfies them, the problem of existence is reduced to the case of ample Riemann forms.

Reduction to the positive definite case

Let $X = \mathbf{C}^n/\Lambda$ be a toroidal group of type q, and let L be a holomorphic line bundle over X which determines the Hermitian form H.

First we consider the *necessary conditions* for $H^0(X, L) \neq 0$.

We have already seen that the condition

(C0) $\quad H_\Lambda = H|_{\mathrm{M}\mathbf{C}_\Lambda \times \mathrm{M}\mathbf{C}_\Lambda}$ is positive semi-definite and not zero

is necessary for $H^0(X, L) \neq 0$ (Theorem 3.2.18).

We shall prove that four other conditions (C1) $\sim$ (C4) are also necessary for $H^0(X, L) \neq 0$.

By Theorem 3.1.4 we may assume that $L = L_\vartheta \otimes L_0$, where L_ϑ is a theta bundle given by a reduced factor ϑ_λ of type (H, ϱ) and L_0 is a topologically trivial line bundle given by an automorphic factor $\iota_\lambda = \mathbf{e}(s_\lambda)$. Consider the condition

(C1) $\mathrm{Ker}(A_\Lambda) \supset \mathrm{Ker}(H_\Lambda)$, where $A_\Lambda = A|_{\mathbf{R}_\Lambda \times \mathbf{R}_\Lambda}$ with $A = \mathrm{Im}H$.

Remark. It is obvious that $\mathrm{MC}_\Lambda \cap \mathrm{Ker}(A_\Lambda) \subset \mathrm{Ker}(H_\Lambda)$. The condition (C1) is satisfied, iff $\mathrm{MC}_\Lambda \cap \mathrm{Ker}(A_\Lambda) = \mathrm{Ker}(H_\Lambda)$.

4.1.1 Theorem (ABE [8])
If $H^0(X,L) \neq 0$, then L satisfies the condition (C1).
Proof
Suppose that $L = L_\vartheta \otimes L_0$ does not satisfy the condition (C1). By Lemma 3.2.15 $H^0(X,L^2)$ generates L^2 on X and L^2 does also not satisfy (C1).
Then we consider the following situation and obtain a contradiction.

$H^0(X,L)$ generates L on X and (C1) is not satisfied. Moreover there exist relatively prime sections $\varphi, \psi \in H^0(X,L)$ (see Remark after Lemma 3.2.15).

Let $\pi : \mathbf{C}^n \longrightarrow X = \mathbf{C}^n/\Lambda$ be the canonical projection. We define a non-constant meromorphic function f on $\mathbf{C}^n$ by

$$f := \frac{\psi \circ \pi}{\varphi \circ \pi}.$$

We set the period group of f

$$P_f := \{a \in \mathbf{C}^n : f(x+a) = f(x) \text{ for all } x \in \mathbf{C}^n\}.$$

Obviously $P_f \supset \Lambda$. Since P_f is a closed subgroup of $\mathbf{C}^n$, we can write $P_f = F \oplus \Gamma$, where F is a complex linear subspace and Γ is a discrete subgroup. Consider the projection $\overline{\tau} : \mathbf{C}^n \longrightarrow \mathbf{C}^n/F$. There exists a meromorphic function g on $\mathbf{C}^n/F$ with period group Γ such that $f = g \circ \overline{\tau}$. Let $\Lambda' := \overline{\tau}(\Lambda)$. Then Λ' is a subgroup of Γ because $\Lambda \subset P_f$. Therefore the projection $\overline{\tau} : \mathbf{C}^n \longrightarrow \mathbf{C}^n/F$ induces an epimorphism $\tau : X \longrightarrow X' := (\mathbf{C}^n/F)/\Gamma$. Since X is toroidal, X' must be toroidal. Let $\pi' : \mathbf{C}^n/F \longrightarrow X'$ be the canonical projection. The function g is locally represented by two relatively prime holomorphic functions. This local representation gives a line bundle L' over X' and relatively prime sections φ', ψ' of L' such that

$$g = \frac{\psi' \circ \pi'}{\varphi' \circ \pi'}.$$

From $f = g \circ \overline{\tau}$ it follows

$$\frac{\psi \circ \pi}{\varphi \circ \pi} = \frac{\psi' \circ \pi' \circ \overline{\tau}}{\varphi' \circ \pi' \circ \overline{\tau}}.$$

Since the pairs (φ, ψ) and (φ', ψ') are both relatively prime pairs, we obtain $L \simeq \tau^* L'$.

Let H' be the Hermitian form determined by $\tau^* L'$. By $L \simeq \tau^* L'$, we have

(1) $$A_\Lambda = A'_\Lambda, \quad \text{where } A' = \mathrm{Im}\, H'.$$

By Theorem 3.2.18 f is constant on $x + \mathrm{Ker}(H_\Lambda)$ for any $x \in \mathbf{C}^n$. Then $\mathrm{Ker}(H_\Lambda) \subset F$. Since f is a non–constant function, MC_Λ is not contained in F. A' is the pull-back of a $\mathbf{R}$-bilinear alternating form on $\mathbf{C}^n/F$. Then we have

$$(2) \qquad A'_\Lambda(x,y) = 0, \quad \text{for all } x \in \mathrm{Ker}(H_\Lambda) \text{ and } y \in \mathbf{R}_\Lambda.$$

On the other hand, the condition (C1) is not satisfied. Then, there exists $x_0 \in \mathrm{Ker}(H_\Lambda)$ but $x_0 \notin \mathrm{Ker}(A_\Lambda)$. We can take $y_0 \in \mathbf{R}_\Lambda$ with $A(x_0, y_0) \neq 0$. By (1) and (2)

$$A(x_0, y_0) = A'(x_0, y_0) = 0.$$

This is a contradiction. *Q.E.D.*

When the condition (C1) is satisfied, we set

(E) $\quad E := \mathrm{Ker}(A_\Lambda) \cup i\, \mathrm{Ker}(A_\Lambda).$

Then E is a complex linear subspace of $\mathbf{C}^n$ with the property $\mathrm{Ker}(H_\Lambda) \subset E$.

4.1.2 Proposition

Let H be a Hermitian form on $\mathbf{C}^n$ satisfying the condition (C0).
Then there exists a positive semi–definite Hermitian form $\tilde{H}$ on $\mathbf{C}^n$ with $\tilde{H} \sim_\Lambda H$, iff H satisfies the condition (C1). Furthermore, in this case we can take $\tilde{H}$ with $\mathrm{Ker}(\tilde{H}) = E$, if H_Λ is not positive definite.

Proof

The proof is a generalization of Lemma 3.1.7.

If H_Λ is positive definite, then this is Lemma 3.1.7 itself. Since $\mathrm{Ker}(H_\Lambda) = 0$, the condition (C1) is satisfied automatically.

We consider the case that H_Λ is positive semi–definite and neither positive definite nor zero.

Suppose that the condition (C1) is satisfied. There exist real linear subspaces V_1 and V_2 of $\mathbf{C}^n$ such that $\mathbf{R}_\Lambda = \mathrm{MC}_\Lambda \oplus V_1 \oplus V_2$, $\mathbf{C}^n = \mathrm{MC}_\Lambda \oplus V_1 \oplus V_2 \oplus iV_1 \oplus iV_2$ and $E = \mathrm{Ker}(H_\Lambda) \oplus V_1 \oplus iV_1$. We note that $\mathrm{Ker}(A_\Lambda) = \mathrm{Ker}(H_\Lambda) \oplus V_1$ in this case. Let $\mathrm{MC}_\Lambda = \mathrm{Ker}(H_\Lambda) \oplus \mathbf{C}^\ell$, $\ell = q - k$, $k = \dim_{\mathbf{C}} \mathrm{Ker}(H_\Lambda)$. Since $H|_{\mathbf{C}^\ell \times \mathbf{C}^\ell} > 0$, there exists $a > 0$ such that

$$H(w', w') = A(w', iw') \geq a\|w'\|^2 \quad \text{for all } w' \in \mathbf{C}^\ell.$$

On the other hand, we have $b > 0$ such that

$$|A(x, iy)| \leq b\,\|x\|\;\|y\| \quad \text{for all } x, y \in \mathbf{C}^n.$$

Take $c > 0$ such as

$$\frac{c-b}{2} - \frac{2b^2}{2} > 0 \quad \text{and} \quad \frac{c-b}{2} - \frac{2b^2}{c-b} > 0.$$

Let $T : V_2 \times V_2 \longrightarrow \mathbf{R}$ be a positive definite $\mathbf{R}$-bilinear symmetric form with

$$T(v_2, v_2) \geq c\|v_2\|^2 \quad \text{for all } v_2 \in V_2.$$

Now, we define an $\mathbf{R}$-bilinear alternating form $A_1 : \mathbf{C}^n \times \mathbf{C}^n \longrightarrow \mathbf{R}$ by

$A_1(x, y) := 0 \quad$ for $x \in \mathbf{MC}_\Lambda$ and $y \in \mathbf{C}^n$,

$A_1(v, x) := 0 \quad$ for $v \in V_1 \oplus V_2$ and $x \in \mathbf{R}_\Lambda$,

$A_1(v_1, iv) := -A(v_1, iv) \quad$ for $v_1 \in V_1$ and $v \in V_1 \oplus V_2$,

$A_1(v_2, iv_1) := A_1(v_1, iv_2) = -A(v_1, iv_2) \quad$ for $v_1 \in V_1$ and $v_2 \in V_2$,

$A_1(v_2, iv_2') := T(v_2, v_2') \quad$ for $v_2, v_2' \in V_2$,

$A_1(iv_1, x) := -A_1(x, iv_1) \quad$ for $v_1 \in V_1$ and $x \in \mathbf{R}_\Lambda$,

$A_1(iv_1, iv) := A_1(v_1, v) = 0 \quad$ for $v_1 \in V_1$ and $v \in V_1 \oplus V_2$,

$A_1(iv_2, x) := -A_1(x, iv_2) \quad$ for $v_2 \in V_2$ and $x \in \mathbf{R}_\Lambda$,

$A_1(iv_2, iv) := A_1(v_2, v) = 0 \quad$ for $v_2 \in V_2$ and $v \in V_1 \oplus V_2$.

Then A_1 has the property

$$A_1(x, y) = A_1(ix, iy) \quad \text{for all } x, y \in \mathbf{C}^n.$$

Therefore there exists a Hermitian form H_1 whose imaginary part is A_1, i.e.

$$H_1(x, y) = A_1(x, iy) + iA_1(x, y).$$

Let $\tilde{H} := H + H_1$. Then $\tilde{H} \sim_\Lambda H$ is obvious for $(A_1)_\Lambda = 0$.
We show $\mathrm{Ker}(\tilde{H}) = E$. Let $\tilde{A} := \mathrm{Im}\, \tilde{H} = A + A_1$. Since $\mathrm{Ker}(\tilde{H}) = \mathrm{Ker}(\tilde{A})$, it suffices to show $\mathrm{Ker}(\tilde{A}) = E$. Noting $E = \mathrm{Ker}(H_\Lambda) \oplus V_1 \oplus iV_1$, we can easily see that $E \subset \mathrm{Ker}(\tilde{A})$. Take any $v_2 \in V_2 \setminus \{0\}$. Since $\mathrm{Ker}(A_\Lambda) \cap V_2 = \{0\}$, there exists $y \in \mathbf{R}_\Lambda$ with $A(v_2, y) \neq 0$. From $(A_1)_\Lambda = 0$ it follows that $\tilde{A}(v_2, y) = A(_2, y) \neq 0$. Then $v_2 \notin \mathrm{Ker}(\tilde{A})$. On the other hand we have $\tilde{A}(iv_2, iy) = \tilde{A}(v_2, y) \neq 0$. Then, $iv_2 \notin \mathrm{Ker}(\tilde{A})$. Therefore $E = \mathrm{Ker}(\tilde{A})$.
Next we show that $\tilde{H}$ is positive definite on $\mathbf{C}^\ell \oplus V_2 \oplus iV_2$. Every element of $\mathbf{C}^\ell \oplus V_2 \oplus iV_2$ is written uniquely as

$$w + u + iv \quad \text{where } w \in \mathbf{C}^\ell \text{ and } u, v \in V_2.$$

We have the following inequalities

$$\tilde{H}(w, w) = H(w, w) \geq a\|w\|^2 \quad \text{for } w \in \mathbf{C}^\ell, \tag{3}$$

$$\left|\tilde{H}(w, u) + \tilde{H}(u, w)\right| = |2A(w, iu)| \leq 2b\,\|w\|\,\|u\| \quad \text{for } w \in \mathbf{C}^\ell \text{ and } u \in V_2, \tag{4}$$

(5) $$\left|\tilde{H}(w,iv)+\tilde{H}(iv,w)\right| \le 2b\,\|w\|\,\|v\| \quad \text{for } w\in \mathbf{C}^\ell \text{ and } v\in V_2,$$

(6) $$\left|\tilde{H}(u,iv)+\tilde{H}(iv,u)\right| \le 2b\,\|u\|\,\|v\| \quad \text{for } u,v\in V_2,$$

(7) $$\tilde{H}(u,u)=\tilde{A}(u,iu)=A(u,iu)+T(u,u)\ge (c-b)\|u\|^2 \quad \text{for } u\in V_2,$$

(8) $$\tilde{H}(iv,iv)=\tilde{A}(iv,i^2v)\ge (c-b)\|v\|^2 \quad \text{for } v\in V_2.$$

Using (3) $\sim$ (8), we obtain for $w+u+iv\neq 0$,

$$\begin{aligned}
&\tilde{H}(w+u+iv,w+u+iv)\\
&\ge a\|w\|^2-2b\,\|w\|\,\|u\|\,\|-2b\|\,\|w\|\,\|v\|\,\|-2b\|\,\|u\|\,\|v\|\\
&\quad +(c-b)\|u\|^2+(c-b)\|v\|^2\\
&=\frac{a}{2}\left(\|w\|-\frac{2b}{a}\|u\|\right)^2+\left(\frac{c-b}{2}-\frac{2b^2}{a}\right)\|u\|^2\\
&\quad +\frac{a}{2}\left(\|w\|-\frac{2b}{a}\|v\|\right)^2+\left(\frac{c-b}{2}-\frac{2b^2}{a}\right)\|v\|^2\\
&\quad +\frac{c-b}{2}\left(\|u\|-\frac{2b}{c-b}\|v\|\right)^2+\left(\frac{c-b}{2}-\frac{2b^2}{c-b}\right)\|v\|^2\\
&>0.
\end{aligned}$$

Conversely we assume that there exists a positive semi–definite Hermitian form $\tilde{H}$ on $\mathbf{C}^n$ with $\tilde{H}\sim_\Lambda H$. Suppose that the condition (C1) is not fulfilled. Then there exists $x_0\in \mathrm{Ker}(H_\Lambda)$ with $x_0\notin \mathrm{Ker}(A_\Lambda)$. Take $y_0\in \mathbf{R}_\Lambda$ such as $A(y_0,x_0)<0$. Let $y:=iy_0+tx_0$ for $t\in\mathbf{R}$. Then

$$\begin{aligned}
\tilde{H}(y,y)&=\tilde{A}(y,iy)\\
&=\tilde{A}(y_0,iy_0)+2t\tilde{A}(y_0,x_0)\\
&=\tilde{A}(y_0,iy_0)+2tA(y_0,x_0).
\end{aligned}$$

If we take $t>0$ sufficiently large, then $\tilde{H}(y,y)<0$, a contradition. *Q.E.D.*

Suppose that the conditions (C0) and (C1) are satisfied. By the above proposition we may assume that H is positive semi–definite on $\mathbf{C}^n$ and $\mathrm{Ker}(H)=E$. In this case $\mathrm{Ker}(A)\cap\mathbf{R}_\Lambda=\mathrm{Ker}(A_\Lambda)$.

4.1.3 Proposition
The following statement holds.

(C2) $\Lambda^*:=\tilde{\sigma}(\Lambda)$ is a discrete subgroup of $\mathbf{C}^n/E$, where $\tilde{\sigma}:\mathbf{C}^n\longrightarrow\mathbf{C}^n/E$ is the projection with $E:=\mathrm{Ker}(A_\Lambda)\cup i\,\mathrm{Ker}(A_\Lambda)$.

Proof
Suppose that Λ^* is not discrete. Let $\overline{\Lambda^*}$ be the closure of Λ^*. Then there exists a positive dimensional real linear subspace S with $S \subset \overline{\Lambda^*}$. Take any $\lambda \in \Lambda$. For any $x \in \mathbf{R}_\Lambda$ with $\tilde{\sigma}(x) \in S$ there exists a sequence $\{\lambda_j\} \subset \Lambda$ such that $\tilde{\sigma}(\lambda_j) \to \tilde{\sigma}(x)$. Since $\mathrm{Ker}(H) = E$, we can take a Hermitian form H_0 on $\mathbf{C}^n/E$ such that $H = H_0 \circ (\tilde{\sigma} \times \tilde{\sigma})$. Then $A = \mathrm{Im}\, H = A_0 \circ (\tilde{\sigma} \times \tilde{\sigma})$, where $A_0 = \mathrm{Im}\, H_0$. We have

$$A(\lambda, x) = A_0(\tilde{\sigma}(\lambda), \tilde{\sigma}(x)) = \lim_{j\to\infty} A_0(\tilde{\sigma}(\lambda), \tilde{\sigma}(\lambda_j)) = \lim_{j\to\infty} A(\lambda, \lambda_j).$$

Since $A(\lambda, \lambda_j) \in \mathbf{Z}$, there exists $k \in \mathbf{Z}$ depending on λ and x such that $A(\lambda, x) = k$. For any $r > 0$, $A(\lambda, rx) = rx$. On the other hand, there exists $k' \in \mathbf{Z}$ such that $A(\lambda, rx) = k'$, by the same argument. Then $k = k' = 0$. Therefore $x \in \mathrm{Ker}(A_\Lambda)$. This contradicts $\mathrm{Ker}(A_\Lambda) \subset E$. *Q.E.D.*

From the projection $\tilde{\sigma} : \mathbf{C}^n \longrightarrow \mathbf{C}^n/E$ we obtain an epimorphism $\sigma : X = \mathbf{C}^n/\Lambda \longrightarrow Y := (\mathbf{C}^n/E)/\Lambda^*$. Here Y is also a toroidal group for X is so.
It is obvious that $\Lambda \cap E$ is a discrete subgroup of E. Then by REMMERT-MORIMOTO's theorem 1.1.5

$$E/(\Lambda \cap E) = \mathbf{C}^{s_1} \times \mathbf{C}^{*s_2} \times X_0,$$

where X_0 is a toroidal group. Let $\vartheta_{E,\lambda}$ and $\iota_{E,\lambda} = \mathbf{e}(s_{E,\lambda})$ be the restrictions of ϑ_λ and $\iota_\lambda = \mathbf{e}(s_\lambda)$ on $\Lambda \cap E$ respectively. Since $E = \mathrm{Ker}(H) = \mathrm{Ker}(A)$, we have

$$\vartheta_{E,\lambda}(x) = \varrho(\lambda) \quad \text{for all } \lambda \in \Lambda \cap E \text{ and } x \in E.$$

For the automorphic factor $\vartheta_{E,\lambda} \cdot \iota_{E,\lambda}$ of $\Lambda \cap E$ on E we get the

4.1.4 Lemma
If $H^0(X, L) \neq 0$, then the automorphic factor $\vartheta_{E,\lambda} \cdot \iota_{E,\lambda}$ is cobordant to 1.
Proof
Consider the above projection $\sigma : X \longrightarrow Y = (\mathbf{C}^n/E)/\Lambda^*$. Since $H^0(X, L) \neq 0$, there exists a fibre $F = \sigma^{-1}(y)$ for $y \in Y$ such that $H^0(F, L|_F) \neq 0$. We know that $F \simeq E/(\Lambda \cap E)$, $L|_F$ is given by $\vartheta_{E,\lambda} \cdot \iota_{E,\lambda}$ and $L|_F$ is topologically trivial. Then $L|_F$ is analytically trivial by Proposition 3.2.22. *Q.E.D.*

Changing the automorphic factor $\vartheta_{E,\lambda} \cdot \iota_{E,\lambda}$, if necessary, we may assume that $\vartheta_{E,\lambda} \cdot \iota_{E,\lambda} = 1$ when $H^0(X, L) \neq 0$, by the virtue of the above lemma. We define a homomorphism $\varrho_0 : \Lambda \longrightarrow \mathbf{S}^1 = \{z \in \mathbf{C} : |z| = 1\}$ by

$$\varrho_0(\lambda) := \begin{cases} \varrho(\lambda) & \text{if } \lambda \in \Lambda \cap E \\ 1 & \text{otherwise.} \end{cases}$$

Now we consider $\iota_\lambda \cdot \varrho_0(\lambda)$ and $\vartheta_\lambda \cdot \varrho_0(\lambda)^{-1}$ as ι_λ and ϑ_λ, respectively. Then we may assume from the first that ϱ satisfies the following condition (C3), if $H^0(X, L) \neq 0$.

(C3) $\varrho(\lambda) = 1$ for all $\lambda \in \Lambda \cap E$.

Then the condition $\vartheta_{E,\lambda} \cdot \iota_{E,\lambda} = 1$ becomes

$$\vartheta_{E,\lambda} = 1 \quad \text{and} \quad \iota_{E,\lambda} = 1. \tag{9}$$

Theorem 3.2.18 is improved as follows, by the trivial modification of the proof according to the facts that $E = \mathrm{Ker}(H)$ and $\mathrm{Ker}(A_\Lambda) = \mathrm{Ker}(H) \cap \mathbf{R}_\Lambda$.

4.1.5 Theorem

Suppose that $H^0(X, L)$ generates L on X or $H^0(X, L) \neq 0$. Then, for any $\varphi_1, \varphi_2 \in H^0(X, L)$ with $\varphi_2 \neq 0$ the meromorphic function $f := \varphi_1/\varphi_2$ is constant on $x + \pi(E)$ for all $x \in X$, where $\pi : \mathbf{C}^n \longrightarrow X = \mathbf{C}^n/\Lambda$ is the natural projection.

Let $f = \varphi/\psi$ be the meromorphic function in the proof of Theorem 4.1.1. By Theorem 4.1.5 we see $P_f \supset E$. The projection $\tau : X \longrightarrow X' = (\mathbf{C}^n/F)/\Gamma$ in the proof of Theorem 4.1.1 is decomposed into $\tau = \tau' \circ \sigma$ by $\sigma : X \longrightarrow Y$ and $\tau' : X \longrightarrow X'$. Letting L' the pull-back of the line bundle over X' by τ', we have $L^2 = \sigma^* L'$. We assume that L' is given by an automorphic factor $\vartheta^*_{\lambda^*} \cdot \iota^*_{\lambda^*}$ for Λ^* on $\mathbf{C}^n/E$, where $\vartheta^*_{\lambda^*}$ is a reduced theta factor of type (H^*, ϱ^*). Since $\vartheta^2_\lambda \cdot \iota^2_\lambda$ and $(\vartheta^*_{\tilde{\sigma}(\lambda)} \circ \tilde{\sigma}) \cdot (\iota^*_{\tilde{\sigma}(\lambda)} \circ \tilde{\sigma})$ are cobordant and $H = H_0 \circ (\tilde{\sigma} \times \tilde{\sigma})$, we can take H^* such as $H^* = 2H_0$. Then there exists a homomorphism $\mu : \Lambda \longrightarrow \mathbf{S}^1$ such that

$$\vartheta^2_\lambda(x) = \mu(\lambda) \cdot \vartheta^*_{\tilde{\sigma}(\lambda)}(\tilde{\sigma}(x))$$

for ϑ^2_λ and $\vartheta^*_{\tilde{\sigma}(\lambda)} \circ \tilde{\sigma}$ are reduced theta factors with the same Hermitian form. Hence ϑ^2_λ and $\mu(\lambda)^{-1} \cdot \iota^*_{\tilde{\sigma}(\lambda)} \circ \tilde{\sigma}$ are cobordant. We can take a homomorphism $m : \Lambda \longrightarrow \mathbf{R}$ such that $\mu(x) = \mathbf{e}(m(\lambda))$. Let $\iota_\lambda(x) = \mathbf{e}(s_\lambda(x))$ and $\iota^*_{\lambda^*}(y) = \mathbf{e}(s^*_{\lambda^*}(y))$. Then $s_\lambda(x)$ and $-(1/2)m(\lambda) + (1/2)s^*_{\tilde{\sigma}(\lambda)}(\tilde{\sigma}(x))$ are cobordant. For all $\lambda \in \Lambda \cap E$ we have

$$s^*_{\tilde{\sigma}(\lambda)}(\tilde{\sigma}(x)) = s^*_0(\tilde{\sigma}(x)) = 1.$$

On the other hand

$$\varrho(\lambda)^2 = \vartheta_\lambda(0)^2 = \mu(\lambda) \cdot \vartheta^*_{\tilde{\sigma}(\lambda)}(\tilde{\sigma}(0)) = \mu(\lambda)$$

for all $\lambda \in \Lambda \cap E$. By (C3) $\varrho(\lambda) = 1$ on $\lambda \in \Lambda \cap E$. Thus

$$\mu(\lambda) = 1 \quad \text{for all } \lambda \in \Lambda \cap E.$$

Therefore we can take $m : \Lambda \longrightarrow \mathbf{R}$ such as $m(\lambda) = 0$ for $\lambda \in \Lambda \cap E$. Then, the automorphic factor $\mathbf{e}\left(-\frac{1}{2}m(\lambda) + \frac{1}{2}s^*_{\tilde{\sigma}(\lambda)} \circ \tilde{\sigma}\right)$ for Λ on $\mathbf{C}^n$ is cobordant to ι_λ, and it is the pull-back of an automorphic factor for Λ^* on $\mathbf{C}^n/E$. Furthermore

$$\mathbf{e}\left(-\frac{1}{2}m(\lambda)+\frac{1}{2}s^*_{\tilde{\sigma}(\lambda)}\circ\tilde{\sigma}\right)=1 \quad \text{for all } \lambda\in\Lambda\cap E.$$

Therefore ι_λ satisfies the following condition:

(C4) There exists an automorphic factor $\iota^*_{\lambda^*}$ for Λ^* on $\mathbf{C}^n/E$ which defines a topologically trivial line bundle over $Y=(\mathbf{C}^n/E)/\Lambda^*$ such that ι_λ and $\iota^*_{\tilde{\sigma}(\lambda)}\circ\tilde{\sigma}$ are cobordant.

Now we discuss the *sufficiency of the conditions.*

Let $L=L_\vartheta\otimes L_0$ be a line bundle over a toroidal group $X=\mathbf{C}^n/\Lambda$. Suppose that L satisfies the conditions (C0)~(C4). Then we have an epimorphism $\sigma: X\longrightarrow Y=(\mathbf{C}^n/E)/\Lambda^*$ onto a toroidal group Y by the projection $\tilde{\sigma}:\mathbf{C}^n\longrightarrow\mathbf{C}^n/E$, where $E=\mathrm{Ker}(A_\Lambda)\cup i\mathrm{Ker}(A_\Lambda)=\mathrm{Ker}(H)$. There exists a positive definite Hermitian form H_0 on $\mathbf{C}^n/E$ such that $H=H_0\circ(\tilde{\sigma}\times\tilde{\sigma})$. We define a homomorphism $\varrho_0:\Lambda^*\longrightarrow\mathbf{S}^1$ by $\varrho_0(\lambda^*):=\varrho(\lambda)$ for some $\lambda\in\Lambda$ with $\lambda^*=\tilde{\sigma}(\lambda)$. By (C3) ϱ_0 is well-defined. Then ϱ_0 is a semi-character of Λ^* associated with $A_0:=\mathrm{Im}\,H_0$. Therefore there exists a reduced theta factor ϑ_{0,λ^*} for Λ^* of type (H_0,ϱ_0) such that $\vartheta_\lambda=\vartheta_{0,\tilde{\sigma}(\lambda)}\circ\tilde{\sigma}$. Let L'_{ϑ_0} be the theta bundle over Y given by ϑ_{0,λ^*}. Then $L_\vartheta=\sigma^*L'_{\vartheta_0}$. Let $\iota^*_{\lambda^*}$ be the automorphic factor for Λ^* in (C4), and let L'_0 be the topologically trivial line bundle over Y given by $\iota^*_{\lambda^*}$. Obviously $L_0\simeq\sigma^*L'_0$. Thus we have

$$L=L_\vartheta\otimes L_0\simeq\sigma^*(L'_{\vartheta_0}\otimes L'_0).$$

4.1.6 Theorem (Abe [8])

Let $X=\mathbf{C}^n/\Lambda$ be a toroidal group, and let $L=L_\vartheta\otimes L_0$ be a line bundle over X. Suppose that L satisfies conditions (C0)~(C4). Then

$$H^0(X,L)\simeq H^0(Y,L'_{\vartheta_0}\otimes L'_0) \quad \text{if } H^0(Y,L'_{\vartheta_0}\otimes L'_0)\neq 0,$$

where Y and $L'_{\vartheta_0}\otimes L'_0$ are the toroidal group and the line bundle over Y defined as above.

Proof

The epimorphism $\sigma: X\longrightarrow Y$ induces the injection $\sigma^*: H^0(Y,L'_{\vartheta_0}\otimes L'_0)\longrightarrow H^0(X,L)$. For any $\varphi\in H^0(Y,L'_{\vartheta_0}\otimes L'_0)$, $\sigma^*\varphi$ is constant on fibres $\sigma^{-1}(y)$ $(y\in Y)$. If there exists a section $\psi\in H^0(X,L)$ which is not constant on $\sigma^{-1}(y)$, then the meromorphic function $f:=\psi/\sigma^*\varphi$ is not constant on $\sigma^{-1}(y)$. This contradicts Theorem 4.1.5. Hence $\sigma^*: H^0(Y,L'_{\vartheta_0}\otimes L'_0)\longrightarrow H^0(X,L)$ is an isomorphism.

Q.E.D.

By the above theorem, the existence problem of holomorphic sections is reduced to the positive definite case. We shall discuss this problem later.

Further properties of meromorphic functions

The following definition for Riemann forms is natural as we showed before.

4.1.7 Definition
Let $X = \mathbf{C}^n/\Lambda$ be a toroidal group. A Hermitian form H on $\mathbf{C}^n$ is called a **Riemann form** for X, if

(1) $A := \operatorname{Im} H$ is $\mathbf{Z}$-valued on $\Lambda \times \Lambda$,

(2) H satisfies the conditions (C0) and (C1).

If $H_\Lambda > 0$ in (1), then H is an ample Riemann form for X.

By Proposition 4.1.2 we may assume that a Riemann form H is positive semi-definite on $\mathbf{C}^n$.

4.1.8 Lemma
Let H_1, H_2 be Riemann forms for a toroidal group $X = \mathbf{C}^n/\Lambda$. Then $H := H_1 + H_2$ is also a Riemann form for X with

$$\operatorname{Ker}(A_\Lambda) = (\operatorname{Ker}((A_1)_\Lambda)) \cap (\operatorname{Ker}((A_2)_\Lambda)).$$

Proof
The conditions (1) in the definition of the Riemann form and (C0) are fulfilled for H. As we note before the lemma, we may assume that H_1, H_2 and H are positive semi-definite on $\mathbf{C}^n$. Then, it is trivial that

$$\operatorname{Ker}(H) = \operatorname{Ker}(H_1) \cap \operatorname{Ker}(H_2).$$

Using the facts $\operatorname{Ker}(A_\Lambda) = \operatorname{Ker}(H) \cap \mathbf{R}_\Lambda$ and $\operatorname{Ker}(H_\Lambda) = \mathbf{MC}_\Lambda \cap \operatorname{Ker}(A_\Lambda)$, we see that the condition (C1) is satisfied for H and

$$\operatorname{Ker}(A_\Lambda) = (\operatorname{Ker}((A_1)_\Lambda)) \cap (\operatorname{Ker}((A_2)_\Lambda)),$$

as it was to been proved. *Q.E.D.*

Grauert and Remmert [37] proved the meromorphic reduction theorem for compact homogeneous complex manifolds. It was extended to non–compact homogeneous manifolds by Huckleberry and Snow [48].

Then we know the existence of the meromorphic reduction for Lie groups. However the proof of the following theorem due to Abe [4] shows more precisely how to get the meromorphic reduction. We mention that another proof of the theorem is obtained by Capocasa and Catanese [20].

4.1.9 Theorem (Meromorphic reduction)
Let $X = \mathbf{C}^n/\Lambda$ be a toroidal group. Then there exists a holomorphic fibration $\varrho : X \longrightarrow X_1$ over a quasi–Abelian variety X_1 with the connected complex Abelian Lie group as fibres, which has the following properties:

1. ϱ is a homomorphism between toroidal groups.
2. ϱ gives the isomorphism $\varrho^* : \mathcal{M}(X_1) \longrightarrow \mathcal{M}(X)$, $f \mapsto f \circ \varrho$.
3. If $\tau : X \longrightarrow Y$ is a homomorphism into a quasi–Abelian variety Y, then there exists the unique homomorphism $\sigma : X_1 \longrightarrow Y$ with $\tau = \sigma \circ \varrho$. This means that such a quasi–Abelian variety X_1 exists uniquely.

X_1 is called the **meromorphic reduction** of X.

Proof

If there exists no Riemann form for X, then for any line bundle $L \longrightarrow X$ not analytically trivial we have $H^0(X, L) = 0$ by the previous results. Hence $\mathcal{M}(X) = \mathbf{C}$. In this case X_1 is the trivial group.

Suppose that there exists a Riemann form H for X. By Proposition 4.1.8 we may assume that H is positive semi–definite on $\mathbf{C}^n$. We denote by $\mathcal{R}$ the set of all Riemann forms for X which are positive semi–definite on $\mathbf{C}^n$. Let

$$E := \bigcap_{H \in \mathcal{R}} \mathrm{Ker}(H).$$

By Lemma 4.1.8 there exists a Riemann form $\tilde{H} \in \mathcal{R}$ such that $E = \mathrm{Ker}(\tilde{H})$. Consider the canonical projection $\tilde{\varrho} : \mathbf{C}^n \longrightarrow \mathbf{C}^n/E$. By Proposition 4.1.3, $\Lambda^* = \tilde{\varrho}(\Lambda)$ is a discrete subgroup of $\mathbf{C}^n/E$. Let $X_1 := (\mathbf{C}^n/E)/\Lambda^*$. Then $\tilde{\varrho}$ induces an epimorphism $\varrho : X \longrightarrow X_1$. Since we have an ample Riemann form H_1 for X_1 with $\tilde{H} = H_1 \circ (\tilde{\varrho} \times \tilde{\varrho})$, X_1 is of course a quasi–Abelian variety. For any Riemann form $H \in \mathcal{R}$, $E \subset \mathrm{Ker}(H)$. Then we can see that $\varrho^* : \mathcal{M}(X_1) \longrightarrow \mathcal{M}(X)$, $f \mapsto f \circ \varrho$ is an isomorphism using Theorem 4.1.5. A connected complex Abelian Lie group $\mathrm{Ker}\ \varrho = E/(E \cap \Lambda) = (E + \Lambda)/\Lambda$ is a closed complex Lie subgroup of X. Then $\varrho : X \longrightarrow X_1 = X/(\mathrm{Ker}\ \varrho)$ is a fibre bundle with fibre $E/(E \cap \Lambda)$ (cf. HIRZEBRUCH [45]).

Next we show the property (3). Let $\tau : X \longrightarrow Y$ be a homomorphism into a quasi–Abelian variety Y, and let $\varrho : X \longrightarrow X_1$ be the fibration given in the above. For any $x_1 \in X_1$, τ is constant on the fibre $\varrho^{-1}(x_1)$. If it is not so, there exist $x, x' \in \varrho^{-1}(x_1)$ with $\tau(x) \neq \tau(x')$. Since Y is meromorphically separable (Main Theorem 3.2.21), we can take $f \in \mathcal{M}(Y)$ such that $f(\tau(x)) \neq f(\tau(x'))$. However $f \circ \tau \in \mathcal{M}(X)$ must be constant on the fibre $\varrho^{-1}(x_1)$, a contradiction. Therefore we can define a holomorphic mapping $\sigma : X_1 \longrightarrow Y$ by $\sigma(x_1) := \tau(x)$ for some x with $\varrho(x) = x_1$. It is obvious that σ is a homomorphism. Let $\sigma' : X_1 \longrightarrow Y$ be another homomorphism with $\tau = \sigma' \circ \varrho$. Then $\sigma \circ \varrho = \sigma' \circ \varrho$. Since $\varrho : X \longrightarrow X_1$ is onto, $\sigma = \sigma'$. *Q.E.D.*

4.1.10 Corollary

Let X be a toroidal group. Then $\mathcal{M}(X) = \mathbf{C}$ iff X has no Riemann form.

Proof

If there exists a Riemann form for X, then the quasi–Abelian variety X_1

in the meromorphic reduction $\varrho : X \longrightarrow X_1$ is positive dimensional. Thus $\mathcal{M}(X) \simeq \mathcal{M}(X_1) \neq \mathbf{C}$.

Conversely, suppose that there exists a non-constant meromorphic function $f \in \mathcal{M}(X)$. Then we can take a line bundle $L \longrightarrow X$ and two sections $\varphi, \psi \in H^0(X, L)$ with $f = \varphi/\psi$. Let H be the Hermitian form determined by L. Since L satisfies the conditions (C0)~(C4), H must be a Riemann form for X. *Q.E.D.*

The meromorphic function fields of toroidal groups are those of quasi-Abelian varieties, but this quasi-Abelian variety can reduce to a point.

It is well-known, that the meromorphic function field of a complex torus group can have any transcendental degree t in the limits $0 \leq t \leq n$ given by a Theorem of CHOW, if $n > 1$, hence the meromorphic reduction can have any such dimension.

KOPFERMANN gave in 1964 [64] an example of a non-compact toroidal group which has only constants as meromorphic functions.

Example: The basis in standard coordinates

$$P = \begin{pmatrix} 1\,0\,0 & i & i\sqrt{2} \\ 0\,1\,0 & i\sqrt{3} & i\sqrt{5} \\ 0\,0\,1 & i\sqrt{7} & i \end{pmatrix}$$

generates a toroidal group $X = \mathbf{C}^3/\Lambda$ on which all meromorphic functions are constant.

Existence of automorphic forms and Lefschetz type theorems

In this section we state recent results of TAKAYAMA on the existence of holomorphic sections and Lefschetz type theorems. His proofs need more general results and technic for weakly 1-complete manifolds. Our purpose is to give the systematical knowledge about toroidal groups. Then we state only results here. We refer to the original papers for proofs.

Let $X = \mathbf{C}^n/\Lambda$ be a toroidal group. Consider a holomorphic line bundle L over X. As we have seen in Theorem 3.2.13, if L is positive, then its Hermitian form H is positive definite on MC_Λ, equivalently on $\mathbf{C}^n$. In other words, there exists a Kähler form in the first Chern class $c_1(L) \in H^2(X, \mathbf{R})$. For compact Kähler manifolds, the converse is shown by virtue of the so-called $\partial\bar{\partial}$-Lemma. Unfortunately, the $\partial\bar{\partial}$-Lemma does not hold in general for toroidal groups. KAZAMA and TAKAYAMA [54] proved that for a toroidal group X the

$\partial\bar{\partial}$-Lemma holds on X, iff X is a toroidal theta group. TAKAYAMA [110] proved the following proposition (Theorem 3.1 in [110]) using the weak $\partial\bar{\partial}$-Lemma (Lemma 3.14 in [110]) instead of the $\partial\bar{\partial}$-Lemma.

4.1.11 Proposition
Let L be a holomorphic line bundle on a toroidal group $X = \mathbf{C}^n/\Lambda$ which determines a Hermitian form H on $\mathbf{C}^n$. Suppose that H is positive definite on MC_Λ. Then, L is positive on any relatively compact open subset of X.

By the above proposition on an ampleness theorem for weakly 1–complete manifolds (Theorem 6.6 in [109]), TAKAYAMA proved the

4.1.12 Theorem (TAKAYAMA [110])
Let L and H be the same as in Proposition 4.1.11 Then the following two conditions are equivalent:
1. L is positive,
2. H is positive definite on MC_Λ.

About the existence of non–trivial sections the following conjecture is well-known:

Conjecture. Let L be a holomorphic line bundle on a toroidal group X which determines a Hermitian form H. If H is positive definite on MC_Λ, then L has a non–trivial section.

Partial results were known by COUSIN [26] and ABE [8]. Recently the mentioned conjecture was proved in the general case by TAKAYAMA.

4.1.13 Theorem (TAKAYAMA [110])
Let L be a holomorphic line bundle on a toroidal group X which determines a Hermitian form H. Suppose that H is positive definite on MC_Λ. Then L has a non–trivial section.
Moreover, the complex linear space $H^0(X, L)$ has the infinite dimension, if X is not compact.

A holomorphic line bundle $L \longrightarrow X$ is said to be **very ample**, if $H^0(X, L)$ gives a holomorphic embedding (i.e. a one-to-one holomorphic immersion) into a complex projective space.
In connection with the existence of sections, TAKAYAMA proved the following Lefschetz type theorem.

4.1.14 Theorem (TAKAYAMA [110])
Let L be a positive line bundle on a toroidal group X. Then L^ℓ is very ample for any integer $\ell \geq 3$.

TAKAYAMA improved the above theorem in the next paper (Theorem 3.4 and Theorem 3.10).

4.1.15 Theorem (TAKAYAMA [111])
Let X be a toroidal group with positive line bundle L. Then

1. L is very ample, if X is torusless,
2. L^2 is very ample, if there does not exist a non–trivial subtorus A of X such that $(A, L|_A)$ is a principally polarized Abelian variety.

The second statement in the above theorem is known as OHBUCHI's Lefschetz Theorem [85] for the compact case.

4.2 Extendable line bundles

A toroidal group has structures of fibre bundles. We can consider extendable line bundles in each compactification of fibrations. We first discuss the case of $\mathbf{C}^{*n-q}$–fibre bundles, and next the case with ample Riemann forms of kind ℓ.

The case of $\mathbf{C}^{*n-q}$–fibre bundles

Let $X = \mathbf{C}^n/\Lambda$ be a non-compact toroidal group of type q. We have seen in 1.1.14 that toroidal coordinates define a representation of X as $\mathbf{C}^{*n-q}$–principal bundle on a q–dimensional torus. The basis P of Λ is written as follows

$$P = \begin{pmatrix} 0 & I_q & \hat{T} \\ I_{n-q} & R_1 & R_2 \end{pmatrix}.$$

The basis $(I_q\ \hat{T})$ gives a q–dimensional torus T. We define a group homomorphism $p : \mathbf{C}^n \longrightarrow \mathbf{C}^q \times \mathbf{C}^{*n-q}$ by

$$p(z_1, \ldots, z_n) := (z_1, \ldots z_q, \mathbf{e}(z_{q+1}), \ldots, \mathbf{e}(z_n)).$$

Then $p(\Lambda)$ is a subgroup of $\mathbf{C}^q \times \mathbf{C}^{*n-q}$, and acts naturally on $\mathbf{C}^q \times \mathbf{C}^{*n-q}$ as a subgroup of automorphisms. Thus we have $X \simeq (\mathbf{C}^q \times \mathbf{C}^{*n-q})/p(\Lambda)$ as complex Lie groups. Any $\eta \in p(\Lambda)$ can be extended to an automorphism

$$\eta : \mathbf{C}^q \times \mathbf{P}_1^{n-q} \longrightarrow \mathbf{C}^q \times \mathbf{P}_1^{n-q},$$

where $\mathbf{P}_1$ is the one–dimensional complex projective space. We write also $p(\Lambda)$ for the group of these extended automorphisms. The action of $p(\Lambda)$ on $\mathbf{C}^q \times \mathbf{P}_1^{n-q}$ is properly discontinuous and fix point free. Then it gives a compact complex manifold $\hat{X} := (\mathbf{C}^q \times \mathbf{P}_1^{n-q})/p(\Lambda)$. Let $\hat{\pi} : \mathbf{C}^q \times \mathbf{P}_1^{n-q} \longrightarrow \hat{X}$ be the canonical projection. We define an embedding $\iota : X \longrightarrow \hat{X}$ through the following commutative diagram

$$\begin{array}{ccc} \mathbf{C}^n & \xrightarrow{\ p\ } & \mathbf{C}^q \times \mathbf{P}_1^{n-q} \\ \pi \big\downarrow & & \big\downarrow \hat{\pi} \\ X & \xrightarrow[\ \iota\ ]{} & \hat{X} \end{array}$$

where $\pi : \mathbf{C}^n \longrightarrow X = \mathbf{C}^n/\Lambda$ is the projection. We see $\iota(X) = \hat{\pi}(\mathbf{C}^q \times \mathbf{C}^{*n-q})$.

Now, consider a holomorphic line bundle L_1 on X. Then $(\iota^{-1})^* L_1$ is a line bundle on $\iota(X)$. The following problem was studied by M. STEIN 1994 in his thesis [108].

Problem. When is there a holomorphic line bundle L on $\hat{X}$ with
$$(\iota^{-1})^* L_1 \simeq L|_{\iota(X)} \ ?$$
The results in this section are due to M. STEIN.

Take a subset $I \subset \{1, \ldots, n-q\}$. Letting $I^c := \{1, \ldots, n-q\} \setminus I$, we set $\mathbf{C}^q \times \mathbf{P}(I) := \mathbf{C}^q \times (X_{q+1}(I) \times \ldots \times X_n(I))$, where

$$X_{q+i}(I) := \begin{cases} \mathbf{C} & \text{if } i \in I \\ \mathbf{P}_1 \setminus \{0\} & \text{if } i \in I^c. \end{cases}$$

Then $\mathbf{C}^q \times \mathbf{P}(I)$ is an open subset of $\mathbf{C}^q \times \mathbf{P}_1^{n-q}$. Letting $X(I) := \hat{\pi}(\mathbf{C}^q \times \mathbf{P}(I))$, we have $\iota(X) \subset X(I) \subset \hat{X}$. We also define $\tilde{X} := \hat{\pi}(\mathbf{C}^q \times \mathbf{P}(\{1, \ldots, n-q\})) = \hat{\pi}(\mathbf{C}^n)$. The projection $\hat{\sigma} : \mathbf{C}^q \times \mathbf{P}_1^{n-q} \longrightarrow \mathbf{C}^q$ onto the space of the first q variables induces the $\mathbf{C}^{*n-q}$-principal bundle $\sigma : \iota(X) \longrightarrow T$ and the $\mathbf{C}^{n-q}$-fibre bundle $\tilde{\sigma} : \tilde{X} \longrightarrow T$.

Consider a holomorphic line bundle $L \longrightarrow X(I)$ with $I \subset \{1, \ldots, n-q\}$. Since $X(I) = \mathbf{C}^q \times \mathbf{P}(I) \simeq \mathbf{C}^n$, L is given by an automorphic factor

$$\alpha : p(\Lambda) \times (\mathbf{C}^q \times \mathbf{P}(I)) \longrightarrow \mathbf{C}^*.$$

The following lemma is obvious.

4.2.1 Lemma
The pull-back $\iota^* L$ of L on X is given by the automorphic factor $\tilde{\alpha} : \Lambda \times \mathbf{C}^n \longrightarrow \mathbf{C}^*$, $\tilde{\alpha}_\lambda(z) := \alpha_{p(\lambda)}(p(z))$.

4.2.2 Lemma
Let $L_1 \longrightarrow X$ be a holomorphic line bundle given by an automorphic factor $\beta : \Lambda \times \mathbf{C}^n \longrightarrow \mathbf{C}^*$. Then there exists a holomorphic line bundle $L \longrightarrow X(I)$ with $(\iota^{-1})^* L_1 \simeq L|_{\iota(X)}$, iff there exist a holomorphic function $\varphi : \mathbf{C}^n \longrightarrow \mathbf{C}^*$ and an automorphic factor $\alpha : p(\Lambda) \times (\mathbf{C}^q \times \mathbf{P}(I)) \longrightarrow \mathbf{C}^*$ such that

$$\varphi(z + \lambda)\beta_\lambda(z)\varphi(z)^{-1} = \alpha_{p(\lambda)}(p(z)) \quad \text{for all } (\lambda, z) \in \Lambda \times \mathbf{C}^n.$$

Proof
If $(\iota^{-1})^* L_1 \simeq L|_{\iota(X)}$, then $L_1 \simeq \iota^*(L|_{\iota(X)})$. Let $\alpha : p(\Lambda) \times (\mathbf{C}^q \times \mathbf{P}(I)) \longrightarrow \mathbf{C}^*$ be the automorphic factor which defines L. Then $\alpha \circ p$ and β are cobordant by Lemma 4.2.1.

The converse is proved by the same lemma. *Q.E.D.*

Let $\beta : \Lambda \times \mathbf{C}^n \longrightarrow \mathbf{C}^*$ be a reduced theta factor of type (H, ϱ), and let L_β be the line bundle on X determined by β. Take the canonical unit vectors $e_1, \ldots, e_n$ of $\mathbf{C}^n$. We consider the following condition:

(C) $\operatorname{Im} H(\lambda, e_j) = 0$ for all $\lambda \in \Lambda$ and $q + 1 \leq j \leq n$.

4.2.3 Lemma
Let $\beta : \Lambda \times \mathbf{C}^n \longrightarrow \mathbf{C}^*$ be a reduced theta factor of type (H, ϱ) which satisfies the condition (C), then for any $I \subset \{1, \ldots, n-q\}$ there exists a holomorphic line bundle $L \longrightarrow X(I)$ with $(\iota^{-1})^* L_\beta \simeq L|_{\iota(X)}$.
Proof
We have the decomposition

$$\mathbf{C}^n = \mathrm{M}\mathbf{C}_\Lambda \oplus V \oplus iV = \mathbf{R}_\Lambda \oplus iV,$$

where $\mathrm{M}\mathbf{C}_\Lambda = \langle e_1, \ldots, e_q \rangle_{\mathbf{C}}$ and $V = \langle e_{q+1}, \ldots, e_n \rangle_{\mathbf{R}}$. By the assumption we have $\mathrm{Im}\, H|_{\mathbf{R}_\Lambda \times V} = 0$. We may assume $\mathrm{Im}\, H|_{\mathbf{C}^n \times V} \equiv 0$ because of the freedom of the choice of $\mathrm{Im}\, H|_{iV \times V}$. Then

$$H(e_j, e_{q+i}) = 0 \quad \text{for } 1 \leq i \leq n-q, 1 \leq j \leq n.$$

This means that H is independent on $z_{q+1}, \ldots, z_n$.
We can write

$$\varrho(e_i) = \mathbf{e}(k(e_i)), \quad k(e_i) \in \mathbf{R} \quad \text{for } i = q+1, \ldots, n.$$

Letting

$$g(z) := -k(e_{q+1}) z_{q+1} - \cdots - k(e_n) z_n,$$

we define

$$\begin{aligned}\beta_{1,\lambda}(z) &:= \mathbf{e}(g(z+\lambda))\beta_\lambda(z)\mathbf{e}(g(z))^{-1} \\ &= \mathbf{e}(g(\lambda))\varrho(\lambda)\mathbf{e}\left(\frac{1}{2i}\left[H(z,\lambda) + \frac{1}{2}H(\lambda,\lambda)\right]\right).\end{aligned}$$

Then β_1 is also a reduced theta factor of type (H, ϱ_1), where $\varrho_1(\lambda) = \mathbf{e}(g(\lambda))\varrho(\lambda)$. From $\varrho_1(e_{q+1}) = \cdots = \varrho_1(e_n) = 1$ it follows

$$\varrho_1(\lambda + a_1 e_{q+1} + \cdots a_{n-q} e_n) = \varrho_1(\lambda) \quad \text{for } \lambda \in \Lambda \text{ and } a_1, \ldots, a_{n-q} \in \mathbf{Z}.$$

Of course β and β_1 are cobordant. Therefore we may assume that ϱ itself has the above property. Then an automorphic form $\alpha : p(\Lambda) \times (\mathbf{C}^q \times \mathbf{C}^{*n-q} \longrightarrow \mathbf{C}^*$ is well-defined by

$$\alpha_{p(\lambda)}(p(z)) := \beta_\lambda(z) \quad \text{for } \lambda \in \Lambda \text{ and } z \in \mathbf{C}^n.$$

Since $\beta_\lambda(z)$ is independent on $z_{q+1}, \ldots, z_n$, α is extendable to $p(\Lambda) \times (\mathbf{C}^q \times \mathbf{P}(I))$. Thus the line bundle L on $X(I)$ defined by α has the required property. *Q.E.D.*

We can see that α is extendable to $p(\Lambda) \times (\mathbf{C}^q \times \mathbf{P}_1^{n-q})$ in the above proof. Then we obtain the

4.2.4 Corollary
If a reduced theta factor β of type (H, ϱ) satisfies the condition (C), then there exists a holomorphic line bundle L on $\hat{X}$ with $(\iota^{-1})^* L_\beta \simeq L|_{\iota(X)}$.

4.2.5 Lemma
Let $\beta : \Lambda \times \mathbf{C}^n \longrightarrow \mathbf{C}^*$ be a reduced theta factor of type (H, ϱ). If there exist a subset $I \subset \{1, \ldots, n-q\}$ and a holomorphic line bundle $L \longrightarrow X(I)$ with $(\iota^{-1})^* L_\beta \simeq L|_{\iota(X)}$, then β satisfies the condition (C).
Proof
Suppose that β does not satisfy the condition (C). Then there exist $\lambda_0 \in \Lambda$ and e_j $(q+1 \leq j \leq n)$ such that Im $H(\lambda_0, e_j) \neq 0$. By Lemma 4.2.2 we have an automorphic factor $\alpha : p(\Lambda) \times (\mathbf{C}^q \times \mathbf{P}(I)) \longrightarrow \mathbf{C}^*$ and a holomorphic function $\varphi : \mathbf{C}^n \longrightarrow \mathbf{C}^*$ such that

$$\alpha_{p(\lambda)}(p(z)) = \varphi(z+\lambda)\beta_\lambda(z)\varphi(z)^{-1} \quad \text{for any } \lambda \in \Lambda \text{ and } z \in \mathbf{C}^n.$$

We set $\tilde{\beta}_\lambda(z) := \varphi(z+\lambda)\beta_\lambda(z)\varphi(z)^{-1}$. Then $\tilde{\beta} = \alpha \circ \varrho$. We can take a mapping $b : \Lambda \times \mathbf{C}^n \longrightarrow \mathbf{C}$ such that $\beta_\lambda(z) = \mathbf{e}(b_\lambda(z))$. Also we have $a : p(\Lambda) \times (\mathbf{C}^q \times \mathbf{P}(I)) \longrightarrow \mathbf{C}$ such that $\alpha_{p(\lambda)}(w) = \mathbf{e}(a_{p(\lambda)}(w))$. If we define $\tilde{b} : \Lambda \times \mathbf{C}^n \longrightarrow \mathbf{C}$ by $\tilde{b} = a \circ p$, then $\tilde{\beta}_\lambda(z) = \mathbf{e}(\tilde{b}_\lambda(z))$. Since $\tilde{b}_\lambda(z+e_j) = \tilde{b}_\lambda(z)$ and $\tilde{b}_{e_j}(z) \in \mathbf{Z}$, we have $\tilde{b}_{\lambda_0}(z+e_j) + \tilde{b}_{e_j}(z) - \tilde{b}_{e_j}(z+\lambda_0) - \tilde{b}_{\lambda_0}(z) = 0$.
On the other hand, we obtain by $L_{\tilde{\beta}} \simeq L_\beta$

$$\begin{aligned} &\tilde{b}_{\lambda_0}(z+e_j) + \tilde{b}_{e_j}(z) - \tilde{b}_{e_j}(z+\lambda_0) - \tilde{b}_{\lambda_0}(z) \\ &\qquad = b_{\lambda_0}(z+e_j) + b_{e_j}(z) - b_{e_j}(z+\lambda_0) - b_{\lambda_0}(z) \\ &\qquad = \text{Im } H(\lambda_0, e_j) \neq 0. \end{aligned}$$

This is a contradiction. *Q.E.D.*

We sum up the above results in the

4.2.6 Theorem
Let $\beta : \Lambda \times \mathbf{C}^n \longrightarrow \mathbf{C}^*$ be a reduced theta factor.
Then there exists a holomorphic line bundle L on $\hat{X}$ (or $\tilde{X}$) with $(\iota^{-1})^* L_\beta \simeq L|_{\iota(X)}$, iff β satisfies the condition (C).

Let L_1 be a holomorphic line bundle on a toroidal group X. Then we have $L_1 \simeq L_\alpha \otimes L_\beta$, where L_α is a topologically trivial line bundle given by an automorphic factor α and L_β is a theta bundle given by a reduced theta factor β of type (H, ϱ). We say that L_1 satisfies the condition (C) if β does it.

The argument in the proof of Lemma 4.2.5 is also valid for a line bundle L_1 with topologically trivial factor L_α. Then we obtain the

4.2.7 Lemma

Let $L_1 \simeq L_\alpha \otimes L_\beta$ be a holomorphic line bundle on a toroidal group X as above. If L_1 does not satisfy the condition (C), then there exists no holomorphic line bundle L on $X(I)$ for any $I \subset \{1, \ldots, n-q\}$ with $(\iota^{-1})^* L_1 \simeq L|_{\iota(X)}$.

4.2.8 Proposition

Let L_1 be a topologically trivial holomorphic line bundle on a toroidal group X. If there exists a line bundle L on $\hat{X}$ with $(\iota^{-1})^* L_1 \simeq L|_{\iota(X)}$, then L_1 is given by a homomorphism $\Lambda \longrightarrow \mathbf{C}^*$.

Proof

We suppose that L_1 is given by an automorphic factor $\alpha_\lambda(z) = \mathbf{e}(a_\lambda(z))$. By Proposition 3.8, we may assume that the automorphic summand $a : \Lambda \times \mathbf{C}^n \longrightarrow \mathbf{C}$ have the following properties:

1) $a_\lambda(z) = a_\lambda(z_{q+1}, \ldots, z_n)$;
2) $a_\lambda(z) = 0$ for all $\lambda = m_1 e_{q+1} + \cdots + m_{n-q} e_n, \ m_1, \cdots, m_{n-q} \in \mathbf{Z}$;
3) $a_\lambda(z)$ is $\mathbf{Z}e_j$-periodic for $j = q+1, \ldots, n$.

By the assumption, for any $I \subset \{1, \ldots, n-q\}$ there exist an automorphic factor $\beta^I : p(\Lambda) \times (\mathbf{C}^q \times \mathbf{P}(I)) \longrightarrow \mathbf{C}^*$ and a holomorphic function $\varphi^I : \mathbf{C}^n \longrightarrow \mathbf{C}^*$ such that

$$(*) \qquad \beta^I_{p(\lambda)}(p(z)) = \varphi^I(z+\lambda)\alpha_\lambda(z)\varphi^I(z)^{-1}.$$

We can take a holomorphic function f^I on $\mathbf{C}^n$ with $\varphi^I(z) = \mathbf{e}(f^I(z))$. Since $\alpha_{e_j}(z) = 1$, $\beta^I_{p(e_j)}(p(z)) = 1$ for $j = q+1, \ldots, n$, $\varphi^I(z)$ is $\mathbf{Z}e_j$-periodic for $j = q+1, \ldots, n$. Then $f^I(z+e_j) - f^I(z) = k_j \in \mathbf{Z}$ $(j = q+1, \ldots, n)$. Therefore $f^I(z) - \sum_{i=q+1}^n k_i z_i$ is $\mathbf{Z}e_j$-periodic $(j = q+1, \ldots, n)$. We write $z = (z', z'') \in \mathbf{C}^q \times \mathbf{C}^{n-q}$. Expanding the periodic part of $f^I(z)$, we obtain

$$f^I(z) = \sum_{\sigma \in \mathbf{Z}^{n-q}} f^I_\sigma(z')\mathbf{e}(\langle \sigma, z'' \rangle) + \sum_{i=q+1}^n k_i z_i.$$

We have also the Fourier expansion of $a_\lambda(z)$

$$a_\lambda(z) = \sum_{\sigma \in \mathbf{Z}^{n-q}} a_{\lambda,\sigma}\mathbf{e}(\langle \sigma, z'' \rangle).$$

Take $b^I : p(\Lambda) \times (\mathbf{C}^q \times \mathbf{P}(I)) \longrightarrow \mathbf{C}$ with $\beta^I = \mathbf{e}(b^I)$. Since $b^I_{p(\lambda)}(p(z))$ is $\mathbf{Z}e_j$-periodic $(j = q+1, \ldots, n)$, we have the similar expansion

$$b^I_{p(\lambda)}(p(z)) = \sum_{\sigma \in \mathbf{Z}^{n-q}} b^I_{\lambda,\sigma}(z')\mathbf{e}(\langle \sigma, z'' \rangle).$$

If $i \in I$, then $X_{q+i}(I) = \mathbf{C}$. Therefore $b^I_{\lambda,\sigma}(z') = 0$ for $\sigma = {}^t(\sigma_1, \ldots, \sigma_{n-q})$ with $\sigma_i < 0$. It follows from $(*)$ that

$$\sum_{\sigma\in\mathbf{Z}^{n-q}} b^I_{\lambda,\sigma}(z')\mathbf{e}(\langle\sigma,z''\rangle)$$
$$\equiv \sum_{\sigma\in\mathbf{Z}^{n-q}} \left(a_{\lambda,\sigma} + f^I_\sigma(z'+\lambda')\mathbf{e}(\langle\sigma,\lambda''\rangle) - f^I_\sigma(z')\right)\mathbf{e}(\langle\sigma,z''\rangle)$$
$$+ \sum_{i=q+1}^{n} k_i\lambda_i \pmod{\mathbf{Z}}.$$

Thus we obtain

$$a_{\lambda,\sigma} + f^I_\sigma(z'+\lambda')\mathbf{e}(\langle\sigma,\lambda''\rangle) - f^I_\sigma(z') = 0$$

for $\sigma = {}^t(\sigma_1,\ldots,\sigma_{n-q})$ with $\sigma_i < 0$. Differentiating the above equality by the variable z_k $(1 \le k \le q)$, we have

$$\frac{\partial f^I_\sigma}{\partial z_k}(z'+\lambda')\mathbf{e}(\langle\sigma,\lambda''\rangle) = \frac{\partial f^I_\sigma}{\partial z_k}(z').$$

We set $(I_q\ \hat{T}) = (t_1,\ldots,t_{2q})$ and $(R_1\ R_2) = (r_1,\ldots,r_{2q})$. Let L_σ be a line bundle on $T = \mathbf{C}^q/(I_q\ \hat{T})\mathbf{Z}^{2q}$ defined by a homomorphism $t_j \mapsto \mathbf{e}(-\langle\sigma,r_j\rangle)$. Then we can consider $\partial f^I_\sigma/\partial z_k$ as a holomorphic section of L_σ. For X is toroidal, L_σ is not analytically trivial (see Remark 1.16). Hence $\partial f^I_\sigma/\partial z_k \equiv 0$ for $k = 1,\ldots,q$. Then we put

$$f^I_\sigma(z') \equiv f^I_\sigma \quad \text{for } \sigma = {}^t(\sigma_1,\ldots,\sigma_{n-q}) \text{ with } \sigma_i < 0.$$

Since $a_{\lambda,\sigma} + f^I_\sigma(\mathbf{e}(\langle\sigma,\lambda''\rangle) - 1) = 0$, we can write

$$f^I_\sigma = -\frac{a_{\lambda,\sigma}}{\mathbf{e}(\langle\sigma,\lambda''\rangle) - 1}$$

for some λ with $\mathbf{e}(\langle\sigma,\lambda''\rangle) \neq 1$. The right-hand side of the above equality does not depend on the choice of such a λ for $a_\lambda(z)$ is an automorphic summand. If $j \in I^c$, we can get by the same argument

$$f^I_\sigma = -\frac{a_{\lambda,\sigma}}{\mathbf{e}(\langle\sigma,\lambda''\rangle) - 1} \quad \text{for } \sigma = {}^t(\sigma_1,\ldots,\sigma_{n-q}) \text{ with } \sigma_i > 0.$$

We set

$$\mathbf{Z}(I) := \{\sigma \in \mathbf{Z}^{n-q};\, \sigma_i < 0 \text{ for some } i \in I \text{ or } \sigma_j > 0 \text{ for some } j \in I^c\}.$$

Then $\sum_{\sigma\in\mathbf{Z}(I)} f^I_\sigma \mathbf{e}(\langle\sigma,z''\rangle)$ is convergent on $\mathbf{C}^{n-q}$. For any $\sigma \in \mathbf{Z}^{n-q}\setminus\{0\}$ we define

$$f_\sigma := -\frac{a_{\lambda,\sigma}}{\mathbf{e}(\langle\sigma,\lambda''\rangle)} \quad \text{for some } \lambda \in \Lambda \text{ with } \mathbf{e}(\langle\sigma,\lambda''\rangle) \neq 1.$$

We note that $\mathbf{Z}^{n-q} \setminus \{0\} = \bigcup_I \mathbf{Z}(I)$ and there exists $I \subset \{1, \dots, n-q\}$ for any $\sigma \in \mathbf{Z}^{n-q} \setminus \{0\}$ such that $\sigma \in \mathbf{Z}(I)$, $f_\sigma = f_\sigma^I$. Then

$$f(z) := \sum_{\sigma \in \mathbf{Z}^{n-q} \setminus \{0\}} f_\sigma \mathbf{e}(\langle \sigma, z'' \rangle)$$

converges on $\mathbf{C}^{n-q}$, and satisfies

$$\begin{aligned} f(z+\lambda) &+ a_\lambda(z) - f(z) \\ &= \sum_{\sigma \in \mathbf{Z}^{n-q} \setminus \{0\}} [a_{\lambda,\sigma} + f_\sigma(\mathbf{e}(\langle \sigma, \lambda'' \rangle) - 1)]\, \mathbf{e}(\langle \sigma, z'' \rangle) + a_{\lambda,0} \\ &= a_{\lambda,0}. \end{aligned}$$

Thus $a_\lambda(z)$ is cobordant to a homomorphism $\Lambda \ni \lambda \mapsto a_{\lambda,0} \in \mathbf{C}$. *Q.E.D.*

4.2.9 Theorem
Let L_1 be a holomorphic line bundle on a toroidal group X. Then there exists a holomorphic line bundle L on $\hat{X}$ with $(\iota^{-1})^* L_1 \simeq L|_{\iota(X)}$, iff L_1 is a theta bundle satisfying the condition (C).
In this case, we can take a theta bundle L' on $T = \mathbf{C}^q/(I_q\ \hat{T})\mathbf{Z}^{2q}$ such that $L_1 \simeq \sigma^* L'$, where $\sigma : X \longrightarrow T$ is a $\mathbf{C}^{*n-q}$-principal bundle.
Proof
Assume that there exists a holomorphic line bundle L on $\hat{X}$ with $(\iota^{-1})^* L_1 \simeq L|_{\iota(X)}$. By Lemma 4.2.7 L_1 satisfies the condition (C). Let $L_1 = L_\alpha \otimes L_\beta$ be the representation as above. We have a holomorphic line bundle $\hat{L}$ on $\hat{X}$ with $(\iota^{-1})^* L_\beta \simeq \hat{L}|_{\iota(X)}$ by Theorem 4.2.5. Then

$$\begin{aligned} (\iota^{-1})^* L_\alpha &= (\iota^{-1})^*(L_1 \otimes L_\beta^{-1}) \\ &\simeq (L \otimes \hat{L}^{-1})|_{\iota(X)}. \end{aligned}$$

By Proposition 4.2.8 α is cobordant to a homomorphism : $\Lambda \longrightarrow \mathbf{C}^*$. Hence L_1 is a theta bundle.

In this case, we may assume that $L_1 = L_\beta$, where β is a reduced theta factor of type (H, ϱ). Furthermore we may assume that H and ϱ have the properties in the proof of Lemma 4.2.3, i.e. H is independent on $z_{q+1}, \dots, z_n$ and

$$\varrho(\lambda + a_1 e_{q+1} + \cdots + a_{n-q} e_n) = \varrho(\lambda)$$

for $\lambda \in \Lambda$ and $a_1, \dots, a_{n-q} \in \mathbf{Z}$. Let $\tilde{\sigma} : \mathbf{C}^n \longrightarrow \mathbf{C}^q$ be the projection onto the space of the first q variables. Then we can define a theta factor $\alpha' : (I_q\ \hat{T})\mathbf{Z}^{2q} \times \mathbf{C}^q \longrightarrow \mathbf{C}^*$ by $\alpha'_{\tilde{\sigma}(\lambda)}(\tilde{\sigma}(z)) := \beta_\lambda(z)$. The theta bundle $L' \longrightarrow T$ defined by α' has the desired property.

The converse is obvious. *Q.E.D.*

The case of kind ℓ

A toroidal group $X = \mathbf{C}^n/\Lambda$ of type q has the structure of the natural $\mathbf{C}^{*n-q}$-principal bundle as seen before. It has of course other fibrations. In this section we consider the extension problem of line bundle concerning the fibration associated with an ample Riemann form of kind ℓ. We must modify some parts of the argument in the previous section. The results in this section are due to ABE and used in [12] to study meromorphic functions admitting an algebraic addition theorem.

Let $X = \mathbf{C}^n/\Lambda$ be a quasi–Abelian variety of type q $(1 \leq q < n)$, and let L_1 be a positive line bundle on it. By Theorem 3.1.4 we have the decomposition $L_1 = L_\vartheta \otimes L_0$, where L_ϑ is defined by a reduced theta factor ϑ_λ of type (H, ϱ) and L_0 is topologically trivial. Then H is an ample Riemann form (Theorem 4.1.12). We may assume that H is positive definite on $\mathbf{C}^n$ (Lemma 3.1.7). By Lemma 3.1.8 and the natural extension of theta factor, there exist a discrete subgroup $\tilde{\Lambda}$ of rank $2n$ and a theta factor $\tilde{\vartheta}_{\tilde{\lambda}}$ on $\tilde{\Lambda} \times \mathbf{C}^n$ such that ϑ_λ is the restriction of $\tilde{\vartheta}_{\tilde{\lambda}}$ on $\Lambda \times \mathbf{C}^n$, $\Lambda \subset \tilde{\Lambda}$ as subgroup and $A = \mathbf{C}^n/\tilde{\Lambda}$ is an Abelian variety.

Now we assume that H is of kind ℓ $(0 \leq 2\ell \leq n - q)$.

We denote by $\mathcal{H}_n$ the Siegel upper half space of degree n.

4.2.10 Normal form

Let $\tilde{P}$ and P be basis of $\tilde{\Lambda}$ and Λ respectively. After a suitable change of basis using invertible matrices and unimodular matrices (see the proof of the Fibration Theorem 3.1.16), we obtain the normal forms of $\tilde{P}$ and P as follows

$$\tilde{P} = (W\ D), \quad W = (w_{ij}) \in \mathcal{H}_n$$

$$D = \begin{pmatrix} d_1 & & \\ & \ddots & \\ & & d_n \end{pmatrix},$$

where d_i $(i = 1, \ldots, n)$ are positive integers with $d_1|d_2|\cdots|d_n$, and

$$P = \begin{pmatrix} & d_1 & & & & \\ W' & & \ddots & & & \\ & & & d_{q+\ell} & & \\ & & & & d_{q+\ell+1} & \\ W'' & & & & & \ddots \\ & & & & & & d_{n-\ell} \\ & & & 0 & & \end{pmatrix} = (P'\ P''),$$

where we put

$$W' := \begin{pmatrix} w_{11} & \cdots & w_{1,q+\ell} \\ \vdots & & \vdots \\ w_{q+\ell,1} & \cdots & w_{q+\ell,q+\ell} \end{pmatrix} \in \mathcal{H}_{q+\ell},$$

$$W'' := \begin{pmatrix} w_{q+\ell+1,1} & \cdots & w_{q+\ell+1,q+\ell} \\ \vdots & & \vdots \\ w_{n1} & \cdots & w_{n,q+\ell} \end{pmatrix}.$$

Then the projection $\tilde{\tau} : \mathbf{C}^n \longrightarrow \mathbf{C}^{q+\ell}$ onto the first $(q+\ell)$-variables gives a $\mathbf{C}^\ell \times \mathbf{C}^{*n-q-2\ell}$–fibre bundle $\tau : X \longrightarrow A$. Compactifying the fibres, we obtain the associated $\mathbf{P}_1^{n-q-\ell}$–fibre bundle $\overline{\tau} : \overline{X} \longrightarrow A$. We shall study the extendability of L_1 to $\overline{X}$.

We first formulate the problem more precisely as in the previous section. We define a group homomorphism $p : \mathbf{C}^n \longrightarrow \mathbf{C}^{q+\ell} \times \mathbf{C}^{*n-q-2\ell} \times \mathbf{C}^\ell$ by

$$p(z_1, \ldots, z_n) = (z_1, \ldots, z_{q+\ell}, \mathbf{e}(z_{q+\ell+1}), \ldots, \mathbf{e}(z_{n-q}), z_{n-q+1}, \ldots, z_n).$$

The subgroup $p(\Lambda)$ acts properly discontinuous on $\mathbf{C}^{q+\ell} \times \mathbf{P}_1^{n-q-\ell}$. Let $\hat{X} := (\mathbf{C}^{q+\ell} \times \mathbf{P}_1^{n-q-\ell})/p(\Lambda)$. Then we have an isomorphism $X \simeq (\mathbf{C}^{q+\ell} \times \mathbf{C}^{*n-q-2\ell} \times \mathbf{C}^\ell)/p(\Lambda)$ which gives the natural embedding $\iota : X \longrightarrow \hat{X}$. Let $\pi : \mathbf{C}^n \longrightarrow X$ and $\hat{\pi} : \mathbf{C}^{q+\ell} \times \mathbf{P}_1^{n-q-\ell} \longrightarrow \hat{X}$ be projections with $\hat{\pi} \circ p = \iota \circ \pi$.

We consider the Problem on p 106 in this situation. First we give the normal form of ϑ_λ. Let $\tilde{\Lambda}$, $\tilde{\vartheta}_{\tilde{\lambda}}$, $\tilde{P}$, P as above. We write the complex coordinates $(z_1, \ldots, z_n)$ of $\mathbf{C}^n$ as $x = (z, w) = (z_1, \ldots, z_{q+\ell}, w_1, \ldots, w_{n-q-\ell})$. The basis $\tilde{P} = (W \ D)$ is considered as a real basis of $\mathbf{C}^n$. Then any $x \in \mathbf{C}^n$ can be represented as

$$x = Wx' + Dx'', \ x', x'' \in \mathbf{R}^n.$$

We assign

$$\mathbf{x} = \begin{pmatrix} x' \\ x'' \end{pmatrix} \in \mathbf{R}^{2n} \quad \text{for } x \in \mathbf{C}^n.$$

It is well-known that $\tilde{\vartheta}_{\tilde{\lambda}}$ is cobordant to the theta factor $\tilde{\vartheta}_{0,\tilde{\lambda}}$

$$\tilde{\vartheta}_{0,\tilde{\lambda}}(x) = \mathbf{e}\left[- {}^t\lambda' x - \frac{1}{2}\, {}^t\lambda' W \lambda'\right]$$

for $\tilde{\lambda} \in \tilde{\Lambda}$ with $\tilde{\lambda} = W\lambda' + D\lambda''$ and $x \in \mathbf{C}^n$ (cf. Theorem 3.1 in [63]). Then ϑ_λ is cobordant to the restriction $\vartheta_{0,\lambda}$ of $\tilde{\vartheta}_{0,\tilde{\lambda}}$ on $\Lambda \times \mathbf{C}^n$. We note that for $\lambda \in \Lambda$ we have

$$\lambda' = \begin{pmatrix} a \\ 0 \end{pmatrix}, \quad \lambda'' = \begin{pmatrix} b \\ 0 \end{pmatrix},$$

where

$$a = \begin{pmatrix} a_1 \\ \vdots \\ a_{q+\ell} \end{pmatrix} \in \mathbf{Z}^{q+\ell},\ b = \begin{pmatrix} b_1 \\ \vdots \\ b_{n-\ell} \end{pmatrix} \in \mathbf{Z}^{n-\ell}.$$

Then we can write $\vartheta_{0,\lambda}$ explicitly as follows

$$\vartheta_{0,\lambda}(x) = \mathbf{e}\left[-\ {}^t az - \frac{1}{2}\ {}^t aW'a\right]$$

for $\lambda = P'a + P''b$ and $x = (z, w) \in \mathbf{C}^n$. Hence we obtain the

4.2.11 Proposition

Let ϑ_λ be a theta factor with Hermitian form H. If H is ample and of kind ℓ, then we can take a basis P of Λ as in the above and ϑ_λ is represented as

$$\text{(N)} \qquad \vartheta_\lambda(x) = \mathbf{e}\left[-\ {}^t az - \frac{1}{2}\ {}^t aW'a\right]$$

for $\lambda = P'a + P''b \in \Lambda$ and $x = (z, w) \in \mathbf{C}^n$.

We say that (N) is the **normal form** of an ample theta factor ϑ_λ.

Take a subset $I \subset \{1, \dots, n - q - 2\ell\}$. We define

$$X_{q+\ell+i}(I) := \begin{cases} \mathbf{C} & \text{if } i \in I \\ \mathbf{P}_1 \setminus \{0\} & \text{if } i \in I^c. \end{cases}$$

We set

$$\mathbf{P}(I) := X_{q+\ell+1}(I) \times \cdots \times X_{n-\ell}(I) \times (\mathbf{P}_1 \setminus \{0\})^\ell,$$

$$\begin{aligned} X(I) &:= \hat{\pi}(\mathbf{C}^{q+\ell} \times \mathbf{P}(I)) \\ &= (\mathbf{C}^{q+\ell} \times \mathbf{P}(I))/p(\Lambda). \end{aligned}$$

Then we have $\iota(X) \subset X(I) \subset \hat{X}$. Let

$$\Lambda^* := \tilde{\tau}(p(\Lambda)) = \tilde{\tau}(\Lambda),$$

where $\tilde{\tau} : \mathbf{C}^{q+\ell} \times \mathbf{P}_1^{n-q-\ell} \longrightarrow \mathbf{C}^{q+\ell}$ is the projection. Then the basis P^* of Λ^* is

$$P^* = \begin{pmatrix} & d_1 & \\ W' & & \ddots & \\ & & & d_{q+\ell} \end{pmatrix} = (W'\ D')$$

and $A = \mathbf{C}^{q+\ell}/\Lambda^*$. We define a theta factor $\vartheta^*_{0,\lambda^*}$ on $\Lambda^* \times \mathbf{C}^{q+\ell}$ by

$$\vartheta^*_{0,\lambda^*} = \mathbf{e}\left[-\ {}^t a^* z - \frac{1}{2}\ {}^t a^* W' a^*\right]$$

for $\lambda^* = W'a^* + D'b^*$ and $z \in \mathbf{C}^{q+\ell}$. Let $L_{\vartheta_0^*} \longrightarrow A$ be the theta bundle defined by $\vartheta^*_{0,\lambda^*}$. Let $\tilde{L} := \overline{\tau}^* L_{\vartheta_0^*}$ be the pull-back of $L_{\vartheta_0^*}$ by $\overline{\tau} : \hat{X} \longrightarrow A$. For any $I \subset \{1, \ldots, n-q-2\ell\}$ we have $\mathbf{C}^{q+\ell} \times \mathbf{P}(I) \simeq \mathbf{C}^n$. Then $\tilde{L}|_{X(I)}$ is given by an automorphic factor

$$\vartheta_I : p(\Lambda) \times (\mathbf{C}^{q+\ell} \times \mathbf{P}(I)) \longrightarrow \mathbf{C}^*,$$

$$\vartheta_{I,\eta}(y) = \vartheta^*_{0,\overline{\tau}(\eta)}(\overline{\tau}(y)).$$

On the other hand, $\vartheta_\lambda(x) = \vartheta^*_{0,\overline{\tau}\circ p(\lambda)}(\overline{\tau}\circ p(x))$. Hence we have $(\iota^{-1})^* L_\vartheta \simeq \tilde{L}|_{\iota(X)}$. Thus we obtain the following proposition, which says L_ϑ is extendable to the compactification $\hat{X}$.

4.2.12 Proposition
For L_ϑ there exists the line bundle $L_{\vartheta_0^*} \longrightarrow A$ given by a theta factor $\vartheta_0^* : \Lambda^* \times \mathbf{C}^{q+\ell} \longrightarrow \mathbf{C}^*$ such that $L_\vartheta \simeq \tau^* L_{\vartheta_0^*}$, where $\tau : \mathbf{C}^{q+\ell} \times \mathbf{C}^{n-q-\ell} \longrightarrow \mathbf{C}^{q+\ell}$ is the projection.

The rest is the ***investigation of topologically trivial line bundles.***

4.2.13 Proposition
Let $X = \mathbf{C}^n/\Lambda$ be a quasi-Abelian variety as above, and let $L_0 \longrightarrow X$ be a topologically trivial holomorphic line bundle on X. Suppose that there exists a holomorphic line bundle $L \longrightarrow \hat{X}$ such that

$$(\iota^{-1})^* L_0 \simeq L|_{\iota(X)}.$$

Then there exists a homomorphism $\psi : \Lambda^* \longrightarrow \mathbf{C}^*$ such that $L_0 \longrightarrow X$ is given by the homomorphism $\varphi := \psi \circ \tau : \Lambda \longrightarrow \mathbf{C}^*$.

Proof
Let

$$D_{n-\ell} := \begin{pmatrix} d_1 & & \\ & \ddots & \\ & & d_{n-\ell} \end{pmatrix}.$$

We assume that L_0 is defined by an automorphic factor $\alpha_\lambda(x) = \mathbf{e}(a_\lambda(x))$. The automorphic summand $a_\lambda(x)$ has the properties (see Section 2.1):

1. $a_\lambda(x) = 0$ for $\lambda \in D_{n-\ell}\mathbf{Z}^{n-\ell}$;
2. $a_\lambda(x)$ is $D_{n-\ell}\mathbf{Z}^{n-\ell}$–periodic with respect to $x_1, \ldots, x_{n-k}$.

Take a subset $I \subset \{1, \ldots, n-q-2\ell\}$. By Lemma 4.2.2 there exists an automorphic factor $\beta^I : p(\Lambda) \times (\mathbf{C}^{q+\ell} \times \mathbf{P}(I)) \longrightarrow \mathbf{C}^*$ and a holomorphic function $\varphi^I : \mathbf{C}^n \longrightarrow \mathbf{C}^*$ such that

(1) $\beta^I_{p(\lambda)}(p(x)) =$
$$\varphi^I(x+\lambda)\alpha_\lambda(x)\varphi^I(x)^{-1} \quad \text{for all} \quad (\lambda, x) \in (\Lambda \times (\mathbf{C}^n \cap (\mathbf{C}^{q+\ell} \times \mathbf{P}(I))).$$

Let f^I be a holomorphic function on $\mathbf{C}^n$ with $\varphi^I(x) = \mathbf{e}(f^I(x))$. We have an automorphic summand $b^I : p(\Lambda) \times (\mathbf{C}^{q+\ell} \times \mathbf{P}(I)) \longrightarrow \mathbf{C}$ such that $\beta^I_{p(\lambda)}(y) = \mathbf{e}(b^I_{p(\lambda)}(y))$. Let $E = c_1(L_{\beta^I}) \in H^2(X(I), \mathbf{Z})$ represent the first Chern class of the line bundle L_{β^I} on $X(I)$ given by β^I. Then

$$E(p(\lambda_1), p(\lambda_2)) = b^I_{p(\lambda_2)}(p(\lambda_1)y) + b^I_{p(\lambda_1)}(y) - b^I_{p(\lambda_1)}(p(\lambda_2)y) + b^I_{p(\lambda_2)}(y).$$

Since L_0 is topologically trivial, $E(p(\lambda_1), p(\lambda_2)) = 0$ for $\lambda_1, \lambda_2 \in \Lambda$ by (1). Then L_{β^I} is topologically trivial. Therefore we may assume that $\beta^I_{p(d_j e_j)}(p(x)) = 1$ for $j = 1, \ldots, n-\ell$, where e_j is the j-th unit vector of $\mathbf{C}^n$. Hence we have

$$f^I(x + d_j e_j) - f^I(x) = k_j \in \mathbf{Z},\ j = 1, \ldots n-\ell.$$

We define a $D_{n-\ell}\mathbf{Z}^{n-\ell}$-periodic function

$$k^I(x) := f^I(x) - \sum_{j=1}^{n-\ell} \frac{k_j}{d_j} x_j.$$

Let $x = (x', x'', x''') \in \mathbf{C}^{q+\ell} \times \mathbf{C}^{n-q-2\ell} \times \mathbf{C}^{\ell}$. Considering $k^I(x)$ as a periodic function of x'', we obtain the Fourier expansion

$$f^I(x) = \sum_{\sigma \in \mathbf{Z}^{n-q-2\ell}} f^I_\sigma(x', x''')\mathbf{e}(\langle \sigma, x''/d'' \rangle) + \sum_{j=q+\ell+1}^{n-\ell} \frac{k_j}{d_j} x_j,$$

where we set

$$x''/d'' = {}^t(x_{q+\ell+1}/d_{q+\ell+1}, \ldots, x_{n-\ell}/d_{n-\ell}).$$

Similarly we have

$$a_\lambda(x) = \sum_{\sigma \in \mathbf{Z}^{n-q-2\ell}} a_{\lambda,\sigma}(x', x''')\mathbf{e}(\langle \sigma, x''/d'' \rangle),$$

$$b^I_{p(\lambda)}(p(x)) = \sum_{\sigma \in \mathbf{Z}^{n-q-2\ell}} b^I_{\lambda,\sigma}(x', x''')\mathbf{e}(\langle \sigma, x''/d'' \rangle).$$

By the right-hand side of (1), $b^I_{p(\lambda)}(p(x))$ is holomorphically extendable on $\mathbf{C}$ with respect to $x_{n-\ell+j}$ $(j = 1, \ldots, \ell)$.
On the other hand, $b^I_{p(\lambda)}(y)$ is holomorphic on $\mathbf{P}_1 \setminus \{0\}$ with respect to $y_{n-\ell+j}$. Since the $(n-\ell+j)$-th element $(p(x))_{n-\ell+j}$ of $p(x)$ is equal to $x_{n-\ell+j}$, $b^I_{p(\lambda)}(p(x))$ is holomorphic on $\mathbf{P}_1$ with respect to $x_{n-\ell+j}$ $(j = 1, \ldots, \ell)$, hence independent on x'''. Thus we obtain

$$b^I_{p(\lambda)}(p(x)) = \sum_{\sigma \in \mathbf{Z}^{n-q-2\ell}} b^I_{\lambda,\sigma}(x')\mathbf{e}(\langle \sigma, x''/d'' \rangle).$$

For $i \in I$, $X_{q+\ell+i}(I) = \mathbf{C}$. Then $b^I_{p(\lambda)}(y)$ is holomorphic on $\mathbf{C}$ with respect to $y_{q+\ell+i}$. If $j \in I^c$, then $X_{q+\ell+j}(I) = \mathbf{P}_1 \setminus \{0\}$. Therefore $b^I_{p(\lambda)}(y)$ is holomorphic on $\mathbf{P}_1 \setminus \{0\}$ with respect to $y_{q+\ell+j}$. Hence we have

$$(2) \qquad b^I_{\lambda,\sigma}(x') = 0$$

for all $\sigma = {}^t(\sigma_1, \dots, \sigma_{n-q-2\ell})$ with $\sigma_i \langle 0$ for some $i \in I$ or with $\sigma_j > 0$ for some $j \in I^c$. It follows from (1) that

$$(3) \qquad \sum_{\sigma \in \mathbf{Z}^{n-q-2\ell}} b^I_{\lambda,\sigma}(x')\mathbf{e}(\langle \sigma, x''/d''\rangle)$$
$$= \sum_{\sigma \in \mathbf{Z}^{n-q-2\ell}} \big[a_{\lambda,\sigma}(x', x''') + f^I_\sigma(x' + \lambda', x''' + \lambda''')\mathbf{e}(\langle \sigma, \lambda''/d''\rangle)$$
$$-f^I_\sigma(x', x''')\big]\, \mathbf{e}(\langle \sigma, x''/d''\rangle) + \sum_{j=q+\ell+1}^{n-\ell} \frac{k_j}{d_j}\lambda_j + n_\lambda \quad (n_\lambda \in \mathbf{Z}).$$

By (2) and (3) we obtain

$$(4) \qquad a_{\lambda,\sigma}(x', x''') + f^I_\sigma(x' + \lambda', x''' + \lambda''')\mathbf{e}(\langle \sigma, \lambda''/d''\rangle) - f^I_\sigma(x', x''') = 0$$

for all $\sigma = {}^t(\sigma_1, \dots, \sigma_{n-q-2\ell})$ with $\sigma_i < 0$ for some $i \in I$ or with $\sigma_j > 0$ for some $j \in I^c$, and

$$(5) \quad b^I_{\lambda,0}(x') = a_{\lambda,0}(x', x''') + f^I_0(x' + \lambda', x''' + \lambda''') - f^I_0(x', x''') + \text{constant}.$$

We define

$$g_0(x) := f^I_0(x', x'''),$$

$$g^I(x) := \sum_{\substack{\sigma \in \mathbf{Z}^{n-q-2\ell} \\ \sigma_i < 0 \text{ for some } i \in I}} f^I_\sigma(x', x''')\mathbf{e}(\langle \sigma, x''/d''\rangle)$$
$$+ \sum_{\substack{\sigma \in \mathbf{Z}^{n-q-2\ell} \\ \sigma_j > 0 \text{ for some } j \in I^c}} f^I_\sigma(x', x''')\mathbf{e}(\langle \sigma, x''/d''\rangle).$$

Then $g_0(x)$ and $g^I(x)$ are holomorphic functions on $\mathbf{C}^n$.

Take disjoint subsets $I_1, \dots, I_N$ of $\{1, \dots, n - q - 2\ell\}$ with

$$\bigcup_{\alpha=1}^N I_\alpha = \{1, \dots, n - q - 2\ell\}.$$

And we set

$$g(x) := g_0(x) + \sum_{\alpha=1}^{N} g^{I_\alpha}(x).$$

Then $\varphi(x) := \mathbf{e}(g(x))$ is a $D_{n-\ell}\mathbf{Z}^{n-\ell}$-periodic $\mathbf{C}^*$-valued holomorphic function on $\mathbf{C}^n$. We define

$$\tilde{\alpha}_\lambda(x) := \varphi(x+\lambda)\alpha_\lambda(x)\varphi(x)^{-1}.$$

Then $\tilde{\alpha}$ is an automorphic factor cobordant to α. It follows from (4) and (5) that $\tilde{\alpha}_\lambda(x)$ does not depend on (x'', x'''). Therefore there exists an automorphic factor $\alpha_0 : \Lambda^* \times \mathbf{C}^{q+\ell} \longrightarrow \mathbf{C}^*$ such that

$$\tilde{\alpha}_\lambda(x) = \alpha_{0,\bar{\tau}(\lambda)}(\bar{\tau}(x)).$$

The holomorphic line bundle L_{α_0} on $A = \mathbf{C}^{q+\ell}/\Lambda^*$ given by α_0 is topologically trivial, and A is an Abelian variety. Then there exists a homomorphism $\psi : \Lambda^* \longrightarrow \mathbf{C}^*$ such that α_0 and ψ are cobordant. *Q.E.D.*

Combining the above proposition with Proposition 4.2.12, we obtain the

4.2.14 Theorem
Let $L_1 = L_\vartheta \otimes L_0$ be a holomorphic line bundle on a quasi-Abelian variety $X = \mathbf{C}^n/\Lambda$ of type q, where L_ϑ is determined by a reduced theta factor ϑ with Hermitian form H and L_0 is topologically trivial. We assume H is an ample Riemann form of kind ℓ $(0 \leq 2\ell \leq n-q)$. Let $\overline{\tau} : \hat{X} \longrightarrow A = \mathbf{C}^{q+\ell}/\Lambda^*$, $\tau : X \longrightarrow A$ and $\iota : X \longrightarrow \hat{X}$ be as above. Then there exists a holomorphic line bundle $L \longrightarrow \hat{X}$ such that

$$(\iota^{-1})^* L_1 \simeq L|_{\iota(X)}$$

iff there exists a theta bundle $L' \longrightarrow A$ with

$$L_1 \simeq \tau^* L'.$$

Proof
By Proposition 4.2.12 there exists a theta bundle $L_{\vartheta_0^*} \longrightarrow A$ such that $L_\vartheta \simeq \tau^* L_{\vartheta_0^*}$. Then L_ϑ is extendable to $\hat{X}$. Since $L_0 = L_1 \otimes L_\vartheta^{-1}$, there exists a holomorphic line bundle $\tilde{L}_0 \longrightarrow \hat{X}$ such that

$$(\iota^{-1})^* L_0 \simeq \tilde{L}_0|_{\iota(X)}.$$

Thus we obtain a homomorphism $\psi : \Lambda^* \longrightarrow \mathbf{C}^*$ by Proposition 4.2.13 such that L_0 is given by the homomorphism $\psi \circ \tau : \Lambda \longrightarrow \mathbf{C}^*$. *Q.E.D.*

Explicit representation of automorphic forms

In the previous section we obtained the normal form of an ample theta factor. We shall give the explicit formula of automorphic forms for an ample theta factor as an application of Proposition 4.2.11.

Let $X = \mathbf{C}^n/\Lambda$ be a quasi–Abelian variety of type q $(1 \le q < n)$. Suppose that a Hermitian form H of an ample theta factor ϑ_λ is of kind ℓ $(0 \le 2\ell \le n-q)$. Then we can take a basis P of Λ as

$$P = \begin{pmatrix} & d_1 & & & & \\ W' & & \ddots & & & \\ & & & d_{q+\ell} & & \\ & & & & d_{q+\ell+1} & \\ W'' & & & & & \ddots \\ & & & & & & d_{n-\ell} \\ & & & & 0 & & \end{pmatrix} = (P' \; P''),$$

and ϑ_λ is represented as

$$\vartheta_\lambda(x) = \mathbf{e}\left[-\,{}^t a z - \frac{1}{2}\,{}^t a W' a\right]$$

for $\lambda = P'a + P''b \in \Lambda$ and $x = (z, w) \in \mathbf{C}^n$ (Proposition 4.2.11).

The $\mathbf{C}$-vector space $\mathcal{A}_{\vartheta_\lambda}$ of automorphic forms belonging to ϑ_λ consists of all holomorphic functions f on $\mathbf{C}^n$ with

$$(1) \qquad f(x+\lambda) = \vartheta_\lambda(x) f(x), \quad \lambda \in \Lambda,\ x \in \mathbf{C}^n.$$

Let $f \in \mathcal{A}_{\vartheta_\lambda}$. Since $\vartheta_\lambda(x) = 1$ for

$$\lambda \in \begin{pmatrix} d_1 & & \\ & \ddots & \\ & & d_{n-\ell} \\ & 0 & \end{pmatrix} \mathbf{Z}^{n-\ell},$$

f is $d_i\mathbf{Z}$-periodic with respect to z_i $(i = 1, \ldots, q+\ell)$ and $d_{q+\ell+j}\mathbf{Z}$-periodic with respect to w_j $(j = 1, \ldots, n-q-2\ell)$. We put

$$d := \begin{pmatrix} d_1 & & \\ & \ddots & \\ & & d_{q+\ell} \end{pmatrix}.$$

Then f has the following expansion

$$(2) \qquad f(x) = \sum_{\sigma \in d^{-1}\mathbf{Z}^{q+\ell}} f_\sigma(w)\mathbf{e}(\langle \sigma, z\rangle),$$

where $f_\sigma(w)$ is a holomorphic function of w and $d_{q+\ell+j}\mathbf{Z}$-periodic with respect to w_j for $j = 1, \ldots, n-q-2\ell$. For any $\lambda = P'a + P''b \in \Lambda$ and $x = (z, w) \in \mathbf{C}^n$, the left-hand side of (1) is

$$(3) \qquad \sum_{\sigma \in d^{-1}\mathbf{Z}^{q+\ell}} f_\sigma(w + W''a)\mathbf{e}({}^t W'a)\mathbf{e}(\langle \sigma, z\rangle).$$

And the right-hand side of (1) is

$$(4) \qquad \mathbf{e}\left[-\,{}^t az - \frac{1}{2}\,{}^t aW'a\right] \sum_{\sigma \in d^{-1}\mathbf{Z}^{q+\ell}} f_\sigma(w)\mathbf{e}(\langle \sigma, z\rangle)$$

$$= \sum_{\sigma \in d^{-1}\mathbf{Z}^{q+\ell}} \mathbf{e}\left[-\frac{1}{2}\,{}^t aW'a\right] f_\sigma(w)\mathbf{e}(\langle \sigma - a, z\rangle).$$

Comparing (3) with (4), we obtain

$$f_\sigma(w + W''a)\mathbf{e}({}^t\sigma W'a) = \mathbf{e}\left[-\frac{1}{2}\,{}^t aW'a\right] f_{\sigma+a}(w).$$

Therefore we have

$$(5) \qquad f_{\sigma+a}(w) = \mathbf{e}\left[\,{}^t(\sigma + \frac{1}{2}a)W'a\right] f_\sigma(w + W''a)$$

for $\sigma \in d^{-1}\mathbf{Z}^{q+\ell}$ and $a \in \mathbf{Z}^{q+\ell}$.

Consider a holomorphic function $g(w)$ on $\mathbf{C}^{n-q-\ell}$ satisfying the following condition

(∗) g is $d_{q+\ell+j}\mathbf{Z}$-periodic with respect to w_j for $j = 1, \ldots, n-q-2\ell$.

For any $R > 0$ we set

$$\mathcal{H}_R := \{w \in \mathbf{C}^{n-q-\ell} \ ; \ |\mathrm{Im}\, w_i| \le R \ (i = 1, \ldots, n-q-2\ell) \quad |w_j| \le R \ (j = n-q-2\ell+1, \ldots, n-q-\ell)\}.$$

We define

$$f_a(w) := g(w + W''a) \quad \text{for } a \in \mathbf{Z}^{q+\ell}.$$

If we put

$$M_a(R) := \sup_{w \in \mathcal{H}_R} |f_a(w)|,$$

then $M_a(R) < +\infty$ by (*). We set $|a| := \sum_{i=1}^{q+\ell} |a_i|$ for $a = {}^t(a_1, \ldots, a_{q+\ell}) \in \mathbf{Z}^{q+\ell}$.

We denote by $\mathcal{CF}$ the set of all holomorphic functions g satisfying (*) and

$$\limsup_{\substack{|a| \to +\infty \\ a \in \mathbf{Z}^{q+\ell}}} \sqrt[|a|]{\exp[-\pi\, {}^t a(\operatorname{Im} W')a]\, M_a(R)} = 0 \tag{6}$$

for all $R > 0$.

Obviously $\mathcal{CF}$ is a complex linear space. It is easy to see that the coefficient functions f_σ in (2) belong to $\mathcal{CF}$. Then $\mathcal{CF} \neq \emptyset$ if $\mathcal{A}_{\vartheta_\lambda} \neq \emptyset$. On the other hand, we have the

4.2.15 Lemma

$\mathcal{CF} \neq \emptyset$, in fact, $\mathcal{CF}$ has an uncountable basis.

Proof

For any complex numbers $a_1, \ldots, a_\ell$, we define

$$g(w) := \mathbf{e}\left[\sum_{i=1}^{n-q-2\ell} \frac{1}{d_{q+\ell+i}} w_i + \sum_{j=1}^{\ell} a_j w_{n-q-2\ell+j}\right].$$

Then g satisfies (*). Since

$$W''a = \begin{pmatrix} \sum_{i=1}^{q+\ell} a_i w_{q+\ell+1,i} \\ \vdots \\ \sum_{i=1}^{q+\ell} a_i w_{n,i} \end{pmatrix},$$

we have

$$\begin{aligned} |f_a(w)| &= |g(w + W''a)| \\ &= \exp\left[-2\pi \sum_{\alpha=1}^{n-q-2\ell} \frac{1}{d_{q+\ell+\alpha}} \left(\operatorname{Im} w_\alpha + \sum_{i=1}^{q+\ell} a_i \operatorname{Im} w_{q+\ell+\alpha,i}\right)\right. \\ &\qquad \left. -2\pi \sum_{\beta=1}^{\ell} \operatorname{Im}\left(a_\beta \left(w_{n-q-2\ell+\beta} + \sum_{i=1}^{q+\ell} a_i w_{n-\ell+\beta,i}\right)\right)\right]. \end{aligned}$$

We set

$$A := \max\{|a_\beta|\,; \beta = 1, \ldots, \ell\},$$

$$B := \max\{|w_{q+\ell+\alpha,i}|\,; \alpha = 1, \ldots, n-q-\ell \quad \text{and } i = 1, \ldots, q+\ell\}.$$

Let $R > 0$. Then we obtain the following estimates on $\mathcal{H}_R$

$$\left|\sum_{\alpha=1}^{n-q-2\ell} \frac{1}{d_{q+\ell+\alpha}} \left(\operatorname{Im} w_\alpha + \sum_{i=1}^{q+\ell} a_i \operatorname{Im} w_{q+\ell+\alpha,i}\right)\right| \leq (R + B\,|a|) \sum_{\alpha=1}^{n-q-2\ell} \frac{1}{d_{q+\ell+\alpha}},$$

$$\left|\sum_{\beta=1}^{\ell} \mathrm{Im}\left[a_\beta\left(w_{n-q-2\ell+\beta}+\sum_{i=1}^{q+\ell} a_i w_{n-\ell+\beta,i}\right)\right]\right| \le \ell A(R+B\,|a|).$$

Therefore we have

$$M_a(R) \le \exp\left[2\pi(R+B\,|a|)\left(\sum_{\alpha=1}^{n-q-2\ell}\frac{1}{d_{q+\ell+\alpha}}+\ell A\right)\right]. \tag{7}$$

Since Im W' is positive definite, it follows from (7) that

$$\limsup_{\substack{|a|\to+\infty\\ a\in\mathbf{Z}^{q+\ell}}} \sqrt[|a|]{\exp\left[-\pi\, {}^t a(\mathrm{Im}\, W')a\right] M_a(R)} = 0.$$

Hence $g \in \mathcal{CF}$.

We can take arbitrary $a_1, \ldots, a_k \in \mathbf{C}$. Then it is easy to show that $\mathcal{CF}$ has an uncountable basis. *Q.E.D.*

Let $U(d) := \mathrm{Rpv}(d^{-1}\mathbf{Z}^{q+\ell}/\mathbf{Z}^{q+\ell})$ be a complete system of representative of $d^{-1}\mathbf{Z}^{q+\ell}$ mod $\mathbf{Z}^{q+\ell}$. The number of elements $\delta := \sharp U(d)$ of $U(d)$ is given by $\delta = d_1 \cdots d_{q+\ell}$. We set $U(d) = \{s_1, \ldots, s_\delta\}$.

4.2.16 Theorem (Abe [11])

Take any $f_{s_1}, \ldots, f_{s_\delta} \in \mathcal{CF}$. Then the series

$$f(x) := \sum_{i=1}^{\delta}\sum_{a\in\mathbf{Z}^{q+\ell}} f_{s_i+a}(w)\mathbf{e}(\langle s_i+a, z\rangle),\ x=(z,w) \tag{8}$$

converges on $\mathbf{C}^n$ and belongs to $\mathcal{A}_{\vartheta_\lambda}$, where

$$f_{s_i+a}(w) := \mathbf{e}\left[{}^t(s_i+\frac{1}{2}a)W'a\right] f_{s_i}(w+W''a). \tag{9}$$

Conversely, any $f \in \mathcal{A}_{\vartheta_\lambda}$ is represented as in (8). Therefore $\mathcal{A}_{\vartheta_\lambda} \simeq (\mathcal{CF})^\delta$ as complex linear spaces.

Proof

Since

$$\left|\mathbf{e}\left[{}^t(s_i+\frac{1}{2}a)W'a\right]\right| = \exp\left[-2\pi\, {}^t(s_i+\frac{1}{2}a)\mathrm{Im}\, W'a\right],$$

the right-hand side of (8) converges on $\mathbf{C}^n$ by (6). The condition (5) is trivially satisfied by (9). Then $f \in \mathcal{A}_{\vartheta_\lambda}$.

The converse is already seen. *Q.E.D.*

The above formula is for an ample theta factor. If an automorphic factor α_λ with an ample Riemann form H has a wild summand, then it is very difficult to find an explicit formula. For some special type Abe [11] gave it. However, for the general case it is still open.

References

1. ABE, Y. (1987) *(H,C)-groups with positive line bundles.* Nagoya Mathematical Journal **107** 1-1
2. ABE, Y. (1988) *Holomorphic sections of line bundles over (H,C)-groups.* Manuscr. Math. **60** 379-385
3. ABE, Y. (1989) *Homogeneous line bundles over a toroidal group.* Nagoya Mathematical Journal **116** 17-24
4. ABE, Y. (1989)*On toroidal groups.* J. Math. Soc. Japan **41** No. 4, 699-708
5. ABE, Y. (1989) *Another projective embedding of toroidal groups.* Chinese Journal of Mathematics **17** No. 2, 83-86
6. ABE, Y. (1989) *Homomorphisms of toroidal groups.* Mathematics Reports Toyama University **12** 65-112
7. ABE, Y.(1991)*Sur les fonctions périodiques de plusieurs variables.* Nagoya Math. J. **122** 83-114
8. ABE, Y. (1993) *Sur les fonctions périodiques de plusieurs variables II (réduction au cas défini positif).* J. Math. Soc. Japan **45** 59-65
9. ABE, Y. (1994) *Existence of sections of line bundles over a toroidal group and its applications.* Math. Z. **216** 657-664
10. ABE, Y. (1995) *Lefschetz type theorem.* Ann. Mat. Pura Appl. IV. Ser., **169** 1-33
11. ABE, Y. (1997) *Explicit representation of automorphic forms for generalized theta factors.* Unpublished
12. ABE, Y.(1999) *Meromorphic functions admitting an algebraic addition theorem.* Osaka J. Math. **36** 343-363
13. ADAMS, J. F. (1969) *Lectures on Lie Groups.* W. A. Benjamin, New York, 182
14. ANDREOTTI, A. AND GRAUERT, H. (1962) *Théorèmes de finitude pour la cohomologie des espaces complexes.* Bull. Soc. math. Frances **90** 193-259
15. BAKER, A.: *Transcendental Number Theory* (1975) Cambridge University Press
16. BISHOP, R. L. and CRITTENDEN, R. J. (1964) *Geometry of Manifolds.* Academic Press, New York, 273
17. BOTT, R. and CHERN, S.S. (1965) *Hermitian vector bundles and the equidistribution of the zeros of their holomorphic sections.* Acta. Math. **114** 71-112
18. BOURBAKI, N. (1955) *Topologie générale.* Ch. V–VIII, 2. éd., Hermann, Paris
19. CAPOCASA, A. (1987) *Teoria delle funzioni sui quasi-tori.* Diss. di dott., Pisa
20. CAPOCASA, A. and CATANESE, F. (1991) *Periodic meromorphic functions.* Acta. Math. **166** 27-68
21. CAPOCASA, A. and CATANESE, F. (1995) *Linear systems on quasi-abelian vatieties.* Math. Ann. **301** 183-199
22. CHEVALLEY, C. (1957) *Theory of Lie Groups I.* 3rd printing, Princton University Press, 213

23. Conforto, F. (1956) *Abelsche Funktionen und algebraische Geometrie.* Springer, Berlin
24. Cornalba, M. (1976) *Complex tori and Jacobians. In: Complex analysis and its applications.* Intern. Atomic Energy Agency, Vienna, 39-100
25. Cousin, P. (1902) *Sur les fonctions périodiques.* Ann. Sci. École Norm. Sup. **19** 9-61
26. Cousin, P. (1910) *Sur les fonctions triplement périodiques de deux variables.* Acta Math. **33** 105-232
27. Dodson, B. (1980) *Fibrations of quasi-abelian varieties.* Boll. Un. Mat. Ital., Suppl. **2** 283-317
28. Dolbeault, P. (1956) *Formes différentielles et cohomologie sur une variété analytique complexe.* Ann. Math. **64** 83-130
29. Elsner, C.: (1997) Privat information, Hannover
30. Fischer, G. und Forster, O. (1979) *Ein Endlichkeitssatz für Hyperflächen auf kompakten komplexen Räumen.* J. reine angew. Math. **306** 88-93
31. Fujiki, A. (1975) *On the blowing down of analytic spaces.* Publ. RIMS Kyoto Univ. **10** 473-507
32. Gherardelli, F. (1971/73) *Varieté quasi-abeliene (risultati di A. Andreotti e F. Gherardelli).* Seminario di geometria, Univ. Degli Studi, Firenze
33. Gherardelli, F. and Andreotti, A. (1974) *Some remarks on quasi-abelian manifolds. In: Global Analysis and its applications, vol. II.* Intern. Atomic Energy Agency, Vienna, 203-206
34. Grauert, H. (1958) *Analytische Faserungen über holomorph-vollständigen Räumen.* Math. Ann. **135** 263-273
35. Grauert, H. (1963) *Bemerkenswerte pseudokonvexe Mannigfaltigkeiten.* Math. Ann. **81** 377-391
36. Grauert, H. und Fritzsche, K. (1974) *Einführung in die Funktionentheorie mehrerer Veränderlicher.* Springer, Berlin, 213
37. Grauert, H. und Remmert, R. (1962) *Über kompakte homogene komplexe Mannigfaltigkeiten.* Arch. Math., Basel **13**498-507
38. Greenberg, M.J. and Harper, J.R (1981) *Algebraic Topology. A First Course.* W. A. Benjamin Reading, Mass., 311
39. Griffiths, P. and Harris, J. (1978) *Principles of algebraic geometry.* Wiley-Interscience
40. Gröbner, W. (1956) *Matrizenrechnung.* Math. Einzelschriften, Oldenbourg
41. Gunning, R.C. (1956) *The structure of factors of automorphy.* Amer. J. Math. **78** 357-382
42. Hardy, G.H. and Wright, E.M. (1975) *An Introduction to the Theory of Numbers. 5th. ed..* Clarendon Press, Oxford
43. Hefez, A. (1978) *On periodic meromorphic functions on* $\mathbf{C}^n$. Rendiconti della Accad. Naz. dei Lincei, Cl. di Sci. fis. mat. et natur. **64** 255-259
44. Henkin, G.M. and Leiterer, J. (1988) *Andreotti-Grauert Theory by Integral Formulas.* Akademie-Verlag, Berlin **43** 270
45. Hirzebruch, F. (1978) *Topological Methods in Algebraic Geometry. Third edition.* Springer, Berlin, 232
46. Hörmander, L. (1973) *An introduction to complex analysis in several variables.* North Holland Publishing Company, Amsterdam, 208
47. Huckleberry, A.T. and Margulis, G.A. (1983) *Invariant analytic hypersurfaces.* Invent. Math. **71** 235-240

48. HUCKLEBERRY, A.T. and SNOW, D. (1980) *Pseudoconcave homogeneous manifolds.* Ann. Scuola Norm. Sup. Pisa, Cl. Sci. (4) **7** 29-54
49. KAUP, L. (1967) *Eine Künnethformel für Fréchetgarben.* Math. Z. **97** 158-168
50. KAZAMA, H. (1971) *q-complete complex Lie groups and holomorphic fibre bundles over a Stein manifold.* Mém. Fac. Sci. Kyushu Univ., Ser. A **26** 102-118
51. KAZAMA, H. (1973) *On pseudo-convexity of complex abelian Lie groups.* J. Math. Soc. Japan **25** 329-333
52. KAZAMA, H. (1973) *Approximation theorem and application to Nakano's vanishing theorem for weakly 1-complete manifolds.* Mém. Fac. Sci. Kyushu Univ., Ser. A **27** 221-240
53. KAZAMA, H.(1984) *$\overline{\partial}$ Cohomology of (H,C)-groups.* Publ. Res. Inst. Math. Sci. Kyoto Univ. **20** 297-317
54. KAZAMA, H. and TAKAYAMA, S. (1999) *$\partial\overline{\partial}$-Problem on weakly 1-complete Kähler manifolds.* Nagoya Math. J. **155** 81-94
55. KAZAMA, H. and UMENO, T. (1980)*On a Certain Holomorphic Line Bundle over a Compact Non-Kähler Complex Manifold.* Math. Rep. of College of General Educatio, Kyushu Univ. **12** 93-102
56. KAZAMA, H. and UMENO, T. (1984) *Complex abelian Lie groups with finite-dimensional cohomology groups.* J. Math. Soc. Japan **36** 91-106
57. KAZAMA, H. and UMENO, T. (1990) *$\overline{\partial}$ Cohomology of Complex Lie Groups.* Publ. Res. Inst. Math. Sci. Kyoto Univ. **26** 473-484
58. KAZAMA, H. and UMENO, T. (1993) *Dolbeault isomorphisms for holomorphic vector bundles over holomorphic fiber spaces and applications.* J. Math. Soc. Japan **45** No. 1, 121-130
59. KODAIRA, K.(1950) *Differential forms on complex analytic manifolds. Chapter V in "Harmonic Integrals" of G. deRham.* Princeton
60. KODAIRA, K. (1954) *On Kähler varieties of restricted type (an intrinsic characterization of algebraic varieties).* Ann. of Math. **60** 28-48
61. KOIZUMI, S. (1982) *Theta functions* (in Japanese). Kinokuniya, Tokyo
62. KÖNIGSBERGER, K. (1962) *Thetafunktionen und multiplikative automorphe Funktionen zu vorgegebenen Divisoren in komplexen Räumen.* Math. Ann. **148** 147-172
63. KOPFERMANN, K. (1960) *Periodenrelationen ausgearteter komplexer Periodentori.* Math. Ann. **140** 334-343
64. KOPFERMANN, K. (1964) *Maximale Untergruppen Abelscher komplexer Liescher Gruppen.* Schr. Math. Inst. Univ. Münster **29** 72
65. LANG, S. (1972) *Introduction to Algebraic and Abelian Functions.* Addison-Wesley Reading/Mass., 112
66. LANGE, H. and BIRKENHAKE, CH. (1992) *Complex Abelian Varieties.* Springer, Berlin, 435
67. MALGRANGE, B. (1975) *La cohomologie d'une variété analytique complexe à bord pseudo-convex n'est pas nécessairement séparée.* C.R. Acad. Sci. Paris, Sér. A **280** 93-95
68. MATSUSHIMA, Y. (1959) *Fibrés holomorphes sur une tore complexe.* Nagoya Math. J., **14** 1-24
69. MATSUSHIMA, Y. (1972) *Differentiable manifolds.* Marcel Dekker, New York
70. MATSUSHIMA, Y. et MORIMOTO, A. (1960) *Sur certains espaces fibrés holomorphes sur une variété de Stein.* Bull. Soc. math. France **88** 137-155
71. MIZOHATA, S. (1973) *The theory of partial differnential equations.* Camb. Univ. Press, 490

72. MIZUHARA, A. (1971) *On a $\mathbf{P}^1$-bundle over an abelian variety which is almost homogeneous.* Math. Japon. **16** 105-114
73. MORIMOTO, A. (1959) *Sur la classification des espaces fibrés vectories holomorphes sur une tore complexe admettant des connexions holomorphes.* Nagoya Math. J. **15** 83-154
74. MORIMOTO, A. (1965) *Non-compact Complex Lie Groups without Non-constant Holomorphic Functions.* Proceedings of the Conference of Complex Analysis Minneapolis, 256-272
75. MORIMOTO, A. (1966) *On the classification of noncompact complex abelian Lie groups.* Trans. Amer. Math. Soc. **123** 200-228
76. MORROW, J. and KODAIRA, K. (1971) *Complex manifolds.* Rinehart New York, 192
77. MUMFORD, D.(1970) *Abelian Varieties.* Tata Institute of Fundamental Research, Oxford University Press, Bombay, 242
78. NAKANO, S. (1955) *On complex analytic vector bundles.* J. Math. Soc. Japan **7** 1-12
79. NAKANO, S.(1973) *Vanishing theorems for weakly 1-complete manifolds.* Number theory, algebraic geometry and commutative algebra. Kinokuniya,Tokyo
80. NAKANO, S. (1974) *Vanishing theorems for weakly 1-complete manifolds, II.* Publ. Res. Inst. Math. Sci. **10** 101-110
81. NAKANO, S. (1982) *Several complex variables (in Japanese).* Asakura Shoten, Tokyo
82. NAKANO, S. and RHAI, T. (1979) *An attempt towards the global projective embedding of pseudoconvex manifolds.* Chinese J. Math. **7** 47-54
83. NARASIMHAN, R. (1961) *The Levi Problem for Complex Spaces.* Math. Ann. **142** 355-365
84. NARASIMHAN, R. (1963) *The Levi Problem in the theory of functions of several complex variables.* Proc. of the Int. Cong. of Math. 1962, 385-388
85. OHBUCHI, A. (1987) *Some remarks on an ample line bundles on abelian varieties.* Manuscripta Math. 225-238
86. POINCARÉ, H. (1898) *Sur les propriétés du potentiel et sur les fonctions abéliennes. Oeuvres. IV, 162-243.* Acta Mathematica **22** 89-178
87. POTHERING, G. (1977) *Meromorphic function fields of non-compact $\mathbf{C}^n/\Gamma$.* Ph. D. Thesis, Notre Dame, 137
88. REMMERT, R. und VAN DE VEN, A. (1963) *Zur Funktionentheorie homogener komplexer Mannigfaltigkeiten.* Topology **2** 137-157
89. ROSATI, M. (1962) *Le funzioni e le varietà quasi abeliane dalla teoria del Severi ad oggi.* Pontificiae Academiae Scientiarium Scripta Varia. No. 23, Roma, 194
90. ROSENLICHT, M. (1954) *Generalized Jacobian varieties.* Ann. of Math., **59** 505-530
91. ROSENLICHT, M. (1956) *Some basic theorems on algebraic groups.* Amer. J. Math. **78** 401-443
92. ROSENLICHT, M. (1958) *Extensions of vector groups by abelian varieties.* Amer. J. Math. **80** 685-714
93. ROTHSTEIN, W. (1955) *Zur Theorie der analytischen Mannigfaltigkeiten im Raume von n komplexen Veränderlichen.* Math. Ann. **129** 96-
94. ROTHSTEIN, W. und KOPFERMANN, K. (1982) *Funktionentheorie mehrerer komplexer Veränderlicher.* BI-Wissenschaftsverlag, Bibliographisches Institut Mannheim 256

95. SEAGLE, A. S. and R. E. WALDE (1973) *Introduction to Lie groups and Lie Algebras.* Academic Press, New York / London, 361
96. SCHNEIDER, TH. (1957) *Einführung in die transzendenten Zahlen.* Springer, Berlin
97. SELDER, E. (1981) *Lokale analytische Schnittmultiplizitäten.* Diss., Osnabr., 131
98. SELDER, E. (1987) *On the Néron-Severi Group of Abelian Complex Lie Groups.* Ann. Mat. Pura Appl., IV. Ser. **149** 261-285
99. SELDER, E. (1988) *On the Torelli problem of abelian complex Lie groups.* Math. Scand. **62** 173-198
100. SERRE, J.P. (1953) *Quelques problémes globaux relatifs aux variétés de Stein.* Centre Belge Rech. Math., Colloque sur les fonctions de plusieurs variables pp. 57-68. Georges Thone, Liege et Masson & Cie., Paris
101. SERRE, J.P. (1959) *Groupes algebriques et corps de classes.* Hermann, Paris
102. SEVERI, F. (1961) *Funzioni quasi abeliane. Seconda edizione ampliata.* Pontificiae Academiae Scientiarium Scripta Varia. No., 20 Roma, 406
103. SHIFFMAN, B. (1972) *Extension of positive line bundles and meromorphic maps.* Invent. Math. **15** 332-347
104. SIEGEL, C.L. (1948) *Analytic functions of several complex variables.* I.A.S.
105. STEENROD, N. (1951) *The topology of fibre bundles.* Princeton University Press
106. STEIN, K. (1950) *Primfunktionen und multiplikative automorphe Funktionen auf nichtgeschlossenen Riemannschen Flächen und Zylindergebieten.* Acta Math. **83** 165-196
107. STEIN, K. (1951) *Analytische Funktionen mehrerer komplexer Veränderlichen zu vorgegebenen Periodizitätsmoduln und das zweite Cousinsche Problem.* Math. Ann. **123** 201-222
108. STEIN, M. (1994) *Abgeschlossene Untergruppen komplexer abelscher Liescher Gruppen.* Diss., Hannover, 65
109. TAKAYAMA, S. (1998) *Adjoint linear series on weakly 1-complete Kähler manifolds I: global projective embedding.* Math. Ann. **311** 501-531
110. TAKAYAMA, S. (1998) *Adjoint linear series on weakly 1-complete Kähler manifolds II: Lefschetz type theorem on quasi-Abelian varieties.* Math. Ann. **312** 363-385
111. TAKAYAMA, S. (1996) *Adjoint Linear Series on Weakly 1-complete Kähler Manifolds III: Line Bundle Convexity and Very Ampleness.* Preprint, 16
112. TAKEUCHI, S. (1974) *On Completness of holomorphic principal bundles.* Nagoya Math. J., **56** 121-138
113. UMENO, T. (1993) *De Rham cohomology of toroidal groups and Chern classes of complex line bundles.* Pusan Kyŏngnam Math. J., Vol. **9** No. 2, 295-311
114. VARADARAJAN, V.S. (1984) *Lie groups, Lie algebras and their representations.* Prentice-Hall Englewood Cliffs NY, 430
115. VOGT, C. (1981) *Geradenbündel auf toroiden Gruppen.* Diss., Düsseldorf, 111
116. VOGT, C. (1982) *Line bundles on toroidal groups.* J. Reine Angew. Math. **335** 197-215
117. VOGT, C. (1983) *Two remarks concerning toroidal groups.* Manuscripta Mathematica **41** 217-232
118. WEIL, A. (1949) *Théormes fondamentaux de la théorie des fonctions thêta.* Séminaire Bourbaki, Exp. **16**
119. WEIL, A. (1971) *Introduction à l'étude des variétés kählériennes. Nouvelle édition corrigée.* Hermann, Paris
120. WELLS, R.O. (1980) *Differential analysis on complex manifolds, 2nd ed..* Springer, New York

Index

Vol. 1676: P. Cembranos, J. Mendoza, Banach Spaces of Vector-Valued Functions. VIII, 118 pages. 1997.

Vol. 1677: N. Proskurin, Cubic Metaplectic Forms and Theta Functions. VIII, 196 pages. 1998.

Vol. 1678: O. Krupková, The Geometry of Ordinary Variational Equations. X, 251 pages. 1997.

Vol. 1679: K.-G. Grosse-Erdmann, The Blocking Technique. Weighted Mean Operators and Hardy's Inequality. IX, 114 pages. 1998.

Vol. 1680: K.-Z. Li, F. Oort, Moduli of Supersingular Abelian Varieties. V, 116 pages. 1998.

Vol. 1681: G. J. Wirsching, The Dynamical System Generated by the 3n+1 Function. VII, 158 pages. 1998.

Vol. 1682: H.-D. Alber, Materials with Memory. X, 166 pages. 1998.

Vol. 1683: A. Pomp, The Boundary-Domain Integral Method for Elliptic Systems. XVI, 163 pages. 1998.

Vol. 1684: C. A. Berenstein, P. F. Ebenfelt, S. G. Gindikin, S. Helgason, A. E. Tumanov, Integral Geometry, Radon Transforms and Complex Analysis. Firenze, 1996. Editors: E. Casadio Tarabusi, M. A. Picardello, G. Zampieri. VII, 160 pages. 1998

Vol. 1685: S. König, A. Zimmermann, Derived Equivalences for Group Rings. X, 146 pages. 1998.

Vol. 1686: J. Azéma, M. Émery, M. Ledoux, M. Yor (Eds.), Séminaire de Probabilités XXXII. VI, 440 pages. 1998.

Vol. 1687: F. Bornemann, Homogenization in Time of Singularly Perturbed Mechanical Systems. XII, 156 pages. 1998.

Vol. 1688: S. Assing, W. Schmidt, Continuous Strong Markov Processes in Dimension One. XII, 137 page. 1998.

Vol. 1689: W. Fulton, P. Pragacz, Schubert Varieties and Degeneracy Loci. XI, 148 pages. 1998.

Vol. 1690: M. T. Barlow, D. Nualart, Lectures on Probability Theory and Statistics. Editor: P. Bernard. VIII, 237 pages. 1998.

Vol. 1691: R. Bezrukavnikov, M. Finkelberg, V. Schechtman, Factorizable Sheaves and Quantum Groups. X, 282 pages. 1998.

Vol. 1692: T. M. W. Eyre, Quantum Stochastic Calculus and Representations of Lie Superalgebras. IX, 138 pages. 1998.

Vol. 1694: A. Braides, Approximation of Free-Discontinuity Problems. XI, 149 pages. 1998.

Vol. 1695: D. J. Hartfiel, Markov Set-Chains. VIII, 131 pages. 1998.

Vol. 1696: E. Bouscaren (Ed.): Model Theory and Algebraic Geometry. XV, 211 pages. 1998.

Vol. 1697: B. Cockburn, C. Johnson, C.-W. Shu, E. Tadmor, Advanced Numerical Approximation of Nonlinear Hyperbolic Equations. Cetraro, Italy, 1997. Editor: A. Quarteroni. VII, 390 pages. 1998.

Vol. 1698: M. Bhattacharjee, D. Macpherson, R. G. Möller, P. Neumann, Notes on Infinite Permutation Groups. XI, 202 pages. 1998.

Vol. 1699: A. Inoue,Tomita-Takesaki Theory in Algebras of Unbounded Operators. VIII, 241 pages. 1998.

Vol. 1700: W. A. Woyczyński, Burgers-KPZ Turbulence,XI, 318 pages. 1998.

Vol. 1701: Ti-Jun Xiao, J. Liang, The Cauchy Problem of Higher Order Abstract Differential Equations, XII, 302 pages. 1998.

Vol. 1702: J. Ma, J. Yong, Forward-Backward Stochastic Differential Equations and Their Applications. XIII, 270 pages. 1999.

Vol. 1703: R. M. Dudley, R. Norvaiša, Differentiability of Six Operators on Nonsmooth Functions and p-Variation. VIII, 272 pages. 1999.

Vol. 1704: H. Tamanoi, Elliptic Genera and Vertex Operator Super-Algebras. VI, 390 pages. 1999.

Vol. 1705: I. Nikolaev, E. Zhuzhoma, Flows in 2-dimensional Manifolds. XIX, 294 pages. 1999.

Vol. 1706: S. Yu. Pilyugin, Shadowing in Dynamical Systems. XVII, 271 pages. 1999.

Vol. 1707: R. Pytlak, Numerical Methods for Optimal Control Problems with State Constraints. XV, 215 pages. 1999.

Vol. 1708: K. Zuo, Representations of Fundamental Groups of Algebraic Varieties. VII, 139 pages. 1999.

Vol. 1709: J. Azéma, M. Émery, M. Ledoux, M. Yor (Eds), Séminaire de Probabilités XXXIII. VIII, 418 pages. 1999.

Vol. 1710: M. Koecher, The Minnesota Notes on Jordan Algebras and Their Applications. IX, 173 pages. 1999.

Vol. 1711: W. Ricker, Operator Algebras Generated by Commuting Projećtions: A Vector Measure Approach. XVII, 159 pages. 1999.

Vol. 1712: N. Schwartz, J. J. Madden, Semi-algebraic Function Rings and Reflectors of Partially Ordered Rings. XI, 279 pages. 1999.

Vol. 1713: F. Bethuel, G. Huisken, S. Müller, K. Steffen, Calculus of Variations and Geometric Evolution Problems. Cetraro, 1996. Editors: S. Hildebrandt, M. Struwe. VII, 293 pages. 1999.

Vol. 1714: O. Diekmann, R. Durrett, K. P. Hadeler, P. K. Maini, H. L. Smith, Mathematics Inspired by Biology. Martina Franca, 1997. Editors: V. Capasso, O. Diekmann. VII, 268 pages. 1999.

Vol. 1715: N. V. Krylov, M. Röckner, J. Zabczyk, Stochastic PDE's and Kolmogorov Equations in Infinite Dimensions. Cetraro, 1998. Editor: G. Da Prato. VIII, 239 pages. 1999.

Vol. 1716: J. Coates, R. Greenberg, K. A. Ribet, K. Rubin, Arithmetic Theory of Elliptic Curves. Cetraro, 1997. Editor: C. Viola. VIII, 260 pages. 1999.

Vol. 1717: J. Bertoin, F. Martinelli, Y. Peres, Lectures on Probability Theory and Statistics. Saint-Flour, 1997. Editor: P. Bernard. IX, 291 pages. 1999.

Vol. 1718: A. Eberle, Uniqueness and Non-Uniqueness of Semigroups Generated by Singular Diffusion Operators. VIII, 262 pages. 1999.

Vol. 1719: K. R. Meyer, Periodic Solutions of the N-Body Problem. IX, 144 pages. 1999.

Vol. 1720: D. Elworthy, Y. Le Jan, X-M. Li, On the Geometry of Diffusion Operators and Stochastic Flows. IV, 118 pages. 1999.

Vol. 1721: A. Iarrobino, V. Kanev, Power Sums, Gorenstein Algebras, and Determinantal Loci. XXVII, 345 pages. 1999.

Vol. 1722: R. McCutcheon, Elemental Methods in Ergodic Ramsey Theory. VI, 160 pages. 1999.

Vol. 1723: J. P. Croisille, C. Lebeau, Diffraction by an Immersed Elastic Wedge. VI, 134 pages. 1999.

Vol. 1724: V. N. Kolokoltsov, Semiclassical Analysis for Diffusions and Stochastic Processes. VIII, 347 pages. 2000.

Vol. 1725: D. A. Wolf-Gladrow, Lattice-Gas Cellular Automata and Lattice Boltzmann Models. IX, 308 pages. 2000.

Vol. 1726: V. Marić, Regular Variation and Differential Equations. X, 127 pages. 2000.

Vol. 1727: P. Kravanja, M. Van Barel, Computing the Zeros of Analytic Functions. VII, 111 pages. 2000.

Vol. 1728: K. Gatermann, Computer Algebra Methods for Equivariant Dynamical Systems. XV, 153 pages. 2000.

Vol. 1729: J. Azéma, M. Émery, M. Ledoux, M. Yor, Séminaire de Probabilités XXXIV. VI, 431 pages. 2000.

Vol. 1730: S. Graf, H. Luschgy, Foundations of Quantization for Probability Distributions. X, 230 pages. 2000.

Vol. 1731: T. Hsu, Quilts: Central Extensions, Braid Actions, and Finite Groups,. XII, 185 pages. 2000.

Vol. 1732: K. Keller, Invariant Factors, Julia Equivalences and the (Abstract) Mandelbrot Set. X, 206 pages. 2000.

Vol. 1733: K. Ritter, Average-Case Analysis of Numerical Problems. IX, 254 pages. 2000.

Vol. 1734: M. Espedal, A. Fasano, A. Mikelić, Filtration in Porous Media and Industrial Applications. Cetraro 1998. Editor: A. Fasano. 2000.

Vol. 1735: D. Yafaev, Scattering Theory: Some Old and New Problems. XVI, 169 pages. 2000.

Vol. 1736: B. O. Turesson, Nonlinear Potential Theory and Weighted Sobolev Spaces. XIV, 173 pages. 2000.

Vol. 1737: S. Wakabayashi, Classical Microlocal Analysis in the Space of Hyperfunctions. VIII, 367 pages. 2000.

Vol. 1738: M. Émery, A. Nemirovski, D. Voiculescu, Lectures on Probability Theory and Statistics. XI, 356 pages. 2000.

Vol. 1739: R. Burkard, P. Deuflhard, A. Jameson, J.-L. Lions, G. Strang, Computational Mathematics Driven by Industrial Problems. Martina Franca, 1999. Editors: V. Capasso, H. Engl, J. Periaux. VII, 418 pages. 2000.

Vol. 1740: B. Kawohl, O. Pironneau, L. Tartar, J.-P. Zolesio, Optimal Shape Design. Tróia, Portugal 1999. Editors: A. Cellina, A. Ornelas. IX, 388 pages. 2000.

Vol. 1741: E. Lombardi, Oscillatory Integrals and Phenomena Beyond all Algebraic Orders. XV, 413 pages. 2000.

Vol. 1742: A. Unterberger, Quantization and Non-holomorphic Modular Forms.VIII, 253 pages. 2000.

Vol. 1743: L. Habermann, Riemannian Metrics of Constant Mass and Moduli Spaces of Conformal Structures. XII, 116 pages. 2000.

Vol. 1744: M. Kunze, Non-Smooth Dynamical Systems. X, 228 pages. 2000.

Vol. 1745: V. D. Milman, G. Schechtman, Geometric Aspects of Functional Analysis. VIII, 289 pages. 2000.

Vol. 1746: A. Degtyarev, I. Itenberg, V. Kharlamov, Real Enriques Surfaces. XVI, 259 pages. 2000.

Vol. 1747: L. W. Christensen, Gorenstein Dimensions. VIII, 204 pages. 2000.

Vol. 1748: M. Růžička, Electrorheological Fluids: Modeling and Mathematical Theory. XV, 176 pages. 2001.

Vol. 1749: M. Fuchs, G. Seregin, Variational Methods for Problems from Plasticity Theory and for Generalized Newtonian Fluids. VI, 269 pages. 2001.

Vol. 1750: B. Conrad, Grothendieck Duality and Base Change. X, 296 pages. 2001.

Vol. 1751: N. J. Cutland, Loeb Measures in Practice: Recent Advances. XI, 111 pages. 2001.

Vol. 1752: Y. V. Nesterenko, P. Philippon, Introduction to Algebraic Independence Theory. XIII, 256 pages. 2001.

Vol. 1753: A. I. Bobenko, U. Eitner, Painlevé Equations in the Differential Geometry of Surfaces. VI, 120 pages. 2001.

Vol. 1754: W. Bertram, The Geometry of Jordan and Lie Structures. XVI, 269 pages. 2001.

Vol. 1755: J. Azéma, M. Émery, M. Ledoux, M. Yor, Séminaire de Probabilités XXXV. VI, 427 pages. 2001.

Vol. 1756: P. E. Zhidkov, Korteweg de Vries and Nonlinear Schrödinger Equations: Qualitative Theory. VII, 147 pages. 2001.

Vol. 1757: R. R. Phelps, Lectures on Choquet's Theorem. VII, 124 pages. 2001.

Vol. 1758: N. Monod, Continous Bounded Cohomology of Locally Compact Groups. X, 214 pages. 2001.

Vol. 1759: Y. Abe, K. Kopfermann, Toroidal Groups. VIII, 133 pages. 2001.

Vol. 1760: D. Filipović, Consistency Problems for Heath-Jarrow-Morton Interest Rate Models. VIII, 134 pages. 2001.

Vol. 1761: C. Adelman, The Decomposition of Primes in Torsion Point Fields. VI, 142 pages. 2001.

Recent Reprints and New Editions

Vol. 1200: V. D. Milman, G. Schechtman, Asymptotic Theory of Finite Dimensional Normed Spaces - Corrected Second Printing 2001. X, 156 pages. 1986.

Vol. 1618: G. Pisier, Similarity Problems and Completely Bounded Maps - Second, Expanded Edition VII, 198 pages. 2001.

Vol. 1629: J. D. Moore, Lectures on Seiberg-Witten Invariants - Second Edition. VIII, 121 pages. 2001.

Vol. 1702: J. Ma, J. Yong, Forward-Backward Stochastic Differential Equations and Their Applications - Corrected Second Printing 2000. XIII, 270 pages. 1999.

4. Lecture Notes are printed by photo-offset from the master-copy delivered in camera-ready form by the authors. Springer-Verlag provides technical instructions for the preparation of manuscripts. Macro packages in TEX, LATEX2e, LATEX2.09 are available from Springer's web-pages at

http://www.springer.de/math/authors/b-tex.html.

Careful preparation of the manuscripts will help keep production time short and ensure satisfactory appearance of the finished book.

The actual production of a Lecture Notes volume takes approximately 12 weeks.

5. Authors receive a total of 50 free copies of their volume, but no royalties. They are entitled to a discount of 33.3 % on the price of Springer books purchase for their personal use, if ordering directly from Springer-Verlag.

Commitment to publish is made by letter of intent rather than by signing a formal contract. Springer-Verlag secures the copyright for each volume. Authors are free to reuse material contained in their LNM volumes in later publications: A brief written (or e-mail) request for formal permission is sufficient.

Addresses:

Professor J.-M. Morel
CMLA, Ecole Normale Supérieure de Cachan
61 Avenue du Président Wilson
94235 Cachan Cedex France
E-mail: Jean-Michel.Morel@cmla.ens-cachan.fr

Professor B. Teissier
Université Paris 7
UFR de Mathématiques
Equipe Géométrie et Dynamique
Case 7012
2 place Jussieu
75251 Paris Cedex 05
E-mail: Teissier@ens.fr

Professor F. Takens, Mathematisch Instituut,
Rijksuniversiteit Groningen, Postbus 800,
9700 AV Groningen, The Netherlands
E-mail: F.Takens@math.rug.nl

Springer-Verlag, Mathematics Editorial, Tiergartenstr. 17
D-69121 Heidelberg, Germany
Tel.: *49 (6221) 487-701
Fax: *49 (6221) 487-355
E-mail: lnm@Springer.de

9 783540 419891